Digital Control Using Digital Signal Processing

ISBN 0-13-089103-7

90000

9 780130 891037

**Prentice Hall Information
and System Sciences Series**
Thomas Kailath, Editor

Digital Control Using Digital Signal Processing

Farzad Nekoogar

Technical Manager, Intrinsix Corporation
Lecturer, University of California at Berkeley

Gene Moriarty

Professor of Electrical Engineering
San Jose State University, CA

Prentice Hall PTR
Upper Saddle River, New Jersey 07458
http://www.phptr.com

Library of Congress Cataloging-in-Publication Data

Nekoogar, Farzad.
    Digital control using digital signal processing / Farzad Nekoogar,
Gene Moriarty.
        p.   cm.
    Includes bibliographical references and index.
    ISBN 0-13-089103-7
    1. Digital control systems. 2. Signal processing—Digital
techniques. I. Moriarty, Gene. II. Title.
TJ223.M53N45 1998   98-18811
629.8—dc21   CIP

Production supervision: *Jane Bonnell*
Cover design director: *Jerry Votta*
Cover design: *Design Source*
Composition: *Dave Aiken*
Manufacturing manager: *Alan Fischer*
Acquisitions editor: *Bernard M. Goodwin*
Editorial assistant: *Diane Spina*
Marketing manager: *Kaylie Smith*

© 1999 by Prentice Hall PTR
Prentice-Hall, Inc.
A Simon & Schuster Company
Upper Saddle River, New Jersey 07458

Prentice Hall books are widely used by corporations and government agencies for training, marketing, and resale.
The publisher offers discounts on this book when ordered in bulk quantities. For more information, contact Corporate Sales Department, Phone: 800-382-3419; FAX: 201-236-7141;
E-mail: corpsales@prenhall.com
Or write: Prentice Hall PTR, Corporate Sales Dept., One Lake Street, Upper Saddle River, NJ 07458.

MAX5000 and MAX7000 are trademarks of Altera Corporation. MATRIX$_X$ is a registered trademark of Integrated Systems, Inc. MATLAB and SIMULINK are registered trademarks of The MathWorks, Inc. STD 32 is a registered trademark of Ziatech Corporation. Other product and company names mentioned herein are the trademarks or registered trademarks of their respective owners.

Printed in the United States of America
10   9   8   7   6   5   4   3   2

ISBN 0-13-089103-7

Prentice-Hall International (UK) Limited, *London*
Prentice-Hall of Australia Pty. Limited, *Sydney*
Prentice-Hall Canada Inc., *Toronto*
Prentice-Hall Hispanoamericana, S.A., *Mexico*
Prentice-Hall of India Private Limited, *New Delhi*
Prentice-Hall of Japan, Inc., *Tokyo*
Simon & Schuster Asia Pte. Ltd., *Singapore*
Editora Prentice-Hall do Brasil, Ltda., *Rio de Janeiro*

We dedicate this book

To my parents
—Farzad

To Frances, Avelina, and Martin
—Gene

Contents

Preface

This book describes digital control using digital signal processors (DSPs). Most textbooks on digital control contain much control system theory but little information on using DSPs in control systems. Textbooks on DSP cover digital signal processing well but do not show how to use DSPs in control systems. This book covers fundamental digital control theory, as well as DSPs and how to use DSPs in control systems.

This book can be used as part of a first course in digital control systems. It is based on material taught to practicing engineers at the University of California at Berkeley Extension, and on material presented in several controls courses taught to graduate and undergraduate electrical engineering students at San Jose State University. Its level is appropriate for seniors or graduate students in engineering and practicing engineers for self-study. This book requires some background in linear systems theory and some understanding of linear algebra. The first courses in these areas would more than suffice.

The first half of the book (Chapters 1 through 4) covers the basic analysis and design of digital control systems. The second half (Chapters 5 and 6) covers the fundamentals of DSPs as well as modern techniques of digital control system design and compensation using DSPs as discrete controllers.

In Chapter 1 we introduce digital control and contrast it with analog control. There is a brief discussion of the difference between classical control theory and modern control theory. We also present an overview of the design process and the role of DSPs in the design of control systems. The capabilities of CAE packages and their role in the analysis and design of digital control systems are covered. The MATLAB and MATRIX$_x$ software packages, discussed in Appendix A, are tools found to be most useful in the analysis and design of the digital control systems that are at issue in the following chapters.

Chapter 2 covers the basic mathematics of discrete systems: difference equations, the unit pulse response, discrete convolution, z-transforms, the discrete system transfer function, frequency response, Fourier transforms, and mappings from s to z domains. Appendices C and D, on transform pairs and partial-fraction expansions, should be used in conjunction with this chapter.

In Chapter 3 we use discussions of sampled-data systems, state-variable methods, nonlinear systems issues, stability, and sensitivity analysis to give the reader the basic knowledge needed to analyze digital control systems. Appendix E, on matrix algebra, should be reviewed before reading this chapter.

In Chapter 4 we discuss methods used to design control systems and, specifically, controllers used in control systems. We review design parameters from classical control in its discussion of steady-state response and cover conventional classical design methods, such as Bode plots and root locus, and discuss compensation techniques.

Chapter 5 covers the fundamentals of DSPs and shows a general guideline for selecting DSP chips for specific control system applications. The basics of computer architecture as necessary DSP background are provided in Appendix H. We compare analog and digital signal-processing methods and discuss generic DSPs and their architectures. Software and hardware support tools for commercial DSPs are discussed. We include examples of the application of DSPs in control systems.

In Chapter 6 we present the fundamentals of modern control systems design, emphasizing techniques such as state controllability, observability, and pole placement. There is also a brief discussion of the linear quadratic optimal design methodology. The implementation of designs as DSPs is stressed. A detailed example from the area of motion control is included to illustrate these modern design techniques. In addition, the basic ideas of fuzzy logic control are discussed.

In Appendix A we describe the CAE design and analysis packages $MATRIX_x$ and MATLAB, and in Appendix B we describe tools from dSPACE which are used in the implementation of DSP technology in control systems. Appendices C, D, and E provide tables of z-transforms and Laplace transforms, a description of the partial-fraction expansion method, and a description of matrix analysis, respectively. Appendix F contains a functional description of some of the most popular motion controller boards. In Appendix G we give examples of some DSP programs written for control system applications. Appendix H covers the basics of computer architecture.

Undergraduate students should find that the first half of the book, together with selected topics from the final two chapters, provides an adequate introduction to the business of digital control systems, an increasingly important topic of concern in the unfolding of the Information Era. As the century comes to a close, the Age of Energy is shifting into the Age of Information. That disciplines, such as control systems, focused for so long in the analog world, are more and more favoring digital representations and implementations comes as no surprise to most people today. Students, in particular, seem to have no reluctance toward embracing the digitalness of contemporary reality. Graduate students should find sufficient material in the six chapters of the text for a first-year grad course in digital controls using DSPs. They might find that the first three chapters need only be reviewed because most of the material therein has probably been encountered in undergraduate courses.

Instructors might want to skip some material, such as the discussion of discrete equation solutions in the early part of Chapter 2, or they may want to supplement the material in various places, such as the classical frequency-domain topics in Chapter 3, which consider Bode and root-locus analysis but not Nyquist analysis. If students already have a good digital controls background, Chapter 5 on DSP material can be studied independently. Then if the instructor supplements the material with topics from the controls literature, Chapter 5 coupled with Chapter 6 on modern design techniques using DSPs should make a good graduate course at a more advanced level.

If the reader is already a practicing engineer, the book can serve as a reference for digital controls using DSPs, in particular, a theoretical framework for digital controls, the practical aspects with which the reader may be conversant. The use of computer-assisted engineering (CAE) tools should be of special interest to the practicing engineer, who generally needs to handle higher-order systems than most texts deal with.

Acknowledgments

- We are indebted to Dr. Herbert Hanselmann, Chief Executive of dSPACE GmbH, for his suggestions on the book's content.

- We are also indebted to Jack Borninski, of Texas Instruments, Inc., for his detailed review of the manuscript, constructive criticism, suggestions on information to be added, and for his contribution of an important example that we use in the text.

- We also extend our thanks to J. Chris Harvey, our editor, for his editing work, advice on format and writing style, and general knowledge of the publication process.

- We would like to thank the staff of Prentice Hall, Inc., especially Bernard Goodwin and Jane Bonnell, for their support of this project.

- We are also grateful for the fine editing skills of former professor John Lamandella, who helped us gather together many far-flung thoughts.

- Finally, we would like to acknowledge Dave Aiken for his technical input to the book, his computer expertise which was required to pull the book together, and his determination and stamina which were required to complete this project.

In addition, we'd like to thank the following people and companies:

- Mr. Paul Schmidt, of Integrated Systems Technology, Inc., for his assistance.

- Ms. Christina Palumbo, of the MathWorks, Inc.

- F.N.'s students at the University of California at Berkeley Extension, who contributed to the development of the course Digital Control Using Digital Signal Processing and who gave suggestions on this book's content.

- G.M.'s students at San Jose State University, who, during his 20 years of teaching, have contributed directly and indirectly to his understanding of control theory.
- F.N.'s sister, Faranak, and brother, Farhad, for their assistance and support.

F. N.
G. M.

CHAPTER 1

Introduction to Digital Control Using Digital Signal Processing

1.1 BACKGROUND

In this book we introduce the reader to the challenging process of implementing discrete system design concepts by programming digital signal processors (DSPs) to function as controllers in digital control systems. Fig. 1.1 indicates a standard feedback control system configuration. The plant is the process to be controlled. The feedback element is typically a sensor that feeds the plant output back to the input side of the system. The essential task of the designer is to determine the structure of the controller, which is driven by the difference between the input and fed-back output signals.

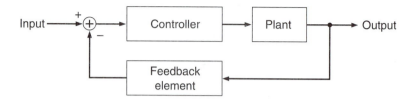

Figure 1.1   Block diagram of a standard feedback control system.

Traditionally, control systems have been designed and analyzed using analog techniques, and their controllers have been implemented with analog components, such as resistors, capacitors, and operational amplifiers. But today, because of the explosive growth and expanding efficiency of digital technology, controllers are typically implemented as programmable digital hardware or as programs on digital computers. The concepts and techniques for analysis and design of digital controllers, implemented as DSPs, are central concerns of this book.

To be useable by digital computers, analog signals need to be sampled and converted to digital form by an analog-to-digital (A/D) converter (also known as an ADC). After being processed by the digital computer, the digital signals need to be converted back to analog form by a digital-to-analog (D/A) converter (DAC). Such a configuration of ADCs, DACs, analog systems to be controlled, sensors, and digital computers functioning as controllers is called a *sampled-data system*. Fig. 1.2 indicates a standard sampled-data digital feedback control system configuration.

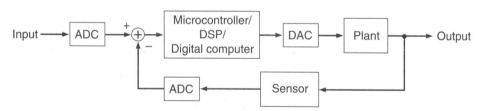

Figure 1.2   Block diagram of a standard digital control system.

Although the expression *digital control system* refers to a control system some part of which is in digital form (and that generally means a sampled-data control system), occasionally we may have a plant completely described by digital mathematics, and when this plant is combined with a digital controller, the system is entirely digital.

Readers may be familiar with the basic ideas of analog control systems analysis and design. Many of the ideas of digital analysis and design transfer over directly from the analog world. Those that do not, of course, require more elaboration. For example, the use of digital computers and microcontrollers such as DSPs *in the loop* for controlling systems is a very different procedure than using computers *outside the loop*, as is common in the analysis and design of both analog and digital controllers. Within the closed loop of a feedback system, a DSP functions as an information-processing device. Similar to the way in which philosopher of technology Albert Borgmann distinguishes between information *about* reality and information *for* reality [1], we might say that a computer in the loop (e.g., a DSP) provides information for the system, whereas a computer outside the loop only provides information about the system.

Although the DSP is an information transformation device, the systems to be controlled, systems such as airplanes or robots or air conditioners, are primarily energy-based devices. But the notions of energy and information are hard to separate. Energy-based systems typically include their information-based controllers, and these controllers, in turn, inform the energy-based systems of which they are a part. At the close of the twentieth century, we are witnessing the rapid proliferation of information-based technology and the gradual diminution of energy-based technology: the postindustrial era is shaping up as the age of information. Yet it seems unlikely that information will totally supplant energy– that virtual reality will totally replace actual reality. The physical, mechanical, and energy aspects of control systems will be essential components for some time to come, because a control system generally seeks to provide or prevent a physical displacement or motion of some kind, and displacement and movement are classical energy notions. Although the hardware or mechanical aspects of energy-based systems tend to become less obtrusive and less obvious to the user, such aspects continue to be of concern to control systems engineers. Despite the recent shift in emphasis between the two, energy and information both play important roles in the control systems of today.

1.2 DIGITAL CONTROL VERSUS ANALOG CONTROL

Fig. 1.3 (elaborated from Fig. 1.1) shows a block diagram representation of a typical analog control system. The basic components of the loop in Fig. 1.3 are the controller, the plant, and the feedback blocks, represented, respectively, by the transfer functions $G_c(s)$, $G_p(s)$, and $H(s)$. Controllers or compensators are usually filters used to compensate or change the frequency response of the system. $R(s)$ is the Laplace transform of the input and $C(s)$ is the Laplace transform of the output.

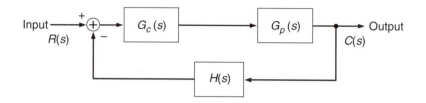

Figure 1.3   Block diagram of an analog control system.

If the output of the summing device is E(s), then

$$\frac{C(s)}{R(s)} = M(s) = \frac{G_c(s)G_p(s)}{1 + G_c(s)G_p(s)H(s)} \tag{1.1}$$

and

$$G_c(s)G_p(s)E(s) = C(s) \tag{1.2}$$

from which

$$\frac{C(s)}{R(s)} = M(s) = \frac{G_c(s)G_p(s)}{1 + G_c(s)G_p(s)H(s)} \tag{1.3}$$

M(s) is called the *closed-loop transfer function.* (We assume that readers are familiar with the Laplace transform and the concept of transfer function as the ratio of Laplace transform of output to Laplace transform of input.)

Fig. 1.4 (elaborated from Fig. 1.2) shows a block diagram of a basic digital control system.

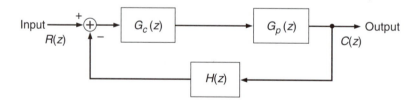

Figure 1.4   Digital control system block diagram.

The controller $G_c(z)$ is usually a microcontroller or a single-chip DSP. The DAC and the ADC converters are essential components, converting data streams from digital information to analog form, and vice versa. DACs and ADCs may or may not be separate discrete elements, but in the block diagram of Fig. 1.4 we assume they are part of the blocks $H(z)$ and $G_c(z)$. (See Chapter 3 for more detailed descriptions of DACs and ADCs.)

If we let $R(z)$ be the z-transform of the input and $C(z)$ the z-transform of the output, then by procedures similar to those employed in the analog case, we can write

$$M(z) = \frac{C(z)}{R(z)} = \frac{G_c(z)G_p(z)}{1 + G_c(z)G_p(z)H(z)} \qquad (1.4)$$

as the closed-loop transfer function for the digital control system of Fig. 1.4. The elements in this expression, including the z-transform theory, are considered in Chapters 2 and 3.

Digital control systems have many advantages over analog systems. They are reliable and very resistant to environmental effects. Because they are capable of time sharing, they can work on multiple tasks. Digital controllers can also be reprogrammed for different tasks, whereas an analog controller, once configured, is usually difficult to modify. Despite these advantages, digital control systems present several problems to their designers. Some problems are related to making continuous signals discrete and selecting sampling rates. Other problems involving single-chip DSPs in control systems are scaling and round-off errors.

Although we have contrasted digital and analog control systems, they do exhibit several similarities. In Chapters 3 and 4, which focus on the analysis and design of digital control systems, we draw on several of the similarities between digital and analog design and analysis techniques.

Example 1.1 Temperature Control System[1]

The objective of the system portrayed in this example is to control temperature in a water tank (Fig. 1.5). Two temperatures are desired, one for the daytime and another for night. This control system contains a water heater tank with one external control input, the user's valve setting, and two resource inputs: feedwater (for filling the tank) and fuel for the heater. There are two measurable external outputs: temperature and water level. Desired water temperatures, different for day and night, are assumed known.

---

[1] This example is presented by courtesy of Integrated Systems, Inc. See Reference [5].

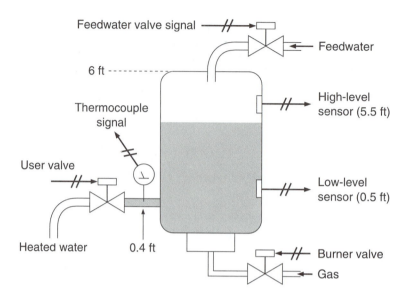

Figure 1.5   Water tank temperature control system.

Points of interest internal to the system include:

- The value of the high-level sensor, typically a voltage that is either high or low, depending on whether the water is above or below a fixed level. The sensor turns on when the water level reaches 5.5 ft.

- The value of the low-level sensor, which is either on or off. It turns on when the water level is above 0.5 ft.

- A thermocouple voltage proportional to water temperature.

- A burner valve control signal, which turns the heater on and off.

- A feedwater valve signal, which lets water into the tank.

    If this system is the plant in either Fig. 1.1 or 1.2, the control engineer's task is to design a controller or compensator, either digital or analog, that processes the outputs – temperature and water level – and produces the best (or at least acceptable) control consisting of plant input and reference signals. This type of control system can be modeled and a controller can be designed using several available computer assisted engineering (CAE) software packages. For example, MATRIX$_x$ (a trademark of Integrated Systems, Inc.) is a control system design and analysis software package that can handle both linear and nonlinear systems as well as both continuous and discrete systems. This CAE package is described briefly in Appendix A.

Example 1.2 Analog and Digital Control of Servo Systems

A closed-loop servo position control system is shown in Fig. 1.6.

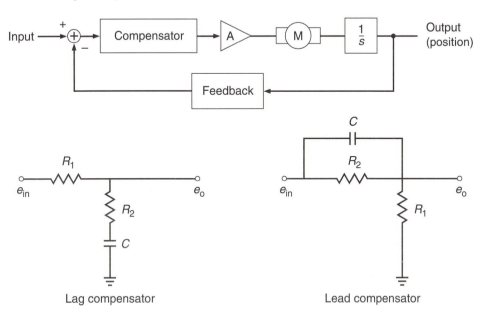

Lag compensator   Lead compensator

Figure 1.6 Analog servo-control system.

The transfer function of its motor is

$$\frac{V_i}{V_m} = \frac{\dfrac{1}{K_e}}{(s\tau_m + 1)(s\tau_e + 1)}$$

where τ_m is the total mechanical time constant and τ_e is the total electrical time constant of the system. The equation for the total mechanical time constant of the system is

$$\tau_m = \frac{RJ}{K_e K_t}$$

where: R is the motor winding resistance, J the total system inertia, K_e the electrical constant of the motor, and K_t the torque constant of the motor.

When an input voltage is applied, the amplifier provides the voltage or current to drive the motor. When the motor shaft reaches the correct position, the feedback element feeds back enough voltage to the summing node to stabilize the system at the final position. But undesired transients in the shaft position may result if the feedback is just a gain term. More sophisticated procedures called *compensation* may be called for. Generally, either a lag or a lead compensator, simple versions of which are shown in Fig. 1.6, can be used to compensate or control this type of system. A lag compensator decreases the bandwidth of the closed loop system and slows the system response. A lead compensator increases the bandwidth and speeds the system response. Compensation is discussed in detail in Chapters 4 and 6.

While the motor in Fig. 1.6 has traditionally been controlled with analog compensators, the more modern trend is toward digital control. Fig. 1.2 illustrates a general digital control system that we can view as a digital servo system if the plant block in the block diagram includes our motor transfer function. The input for this system is a digital word that commands the motor to move to a certain position. The input command is compared to the position data that are fed back by the encoder. If a position error exists, the control equation outputs a digital word, which is converted by the DAC, which drives the motor. As the motor turns, the position count increases or decreases until the error is reduced to zero or the system rests.

In the system shown in Fig. 1.2, the microcontroller replaces the fixed analog compensation block with a control equation, which might take the simple form

$$c(n) = e(n) - kAe(n-1) - Bc(n-1)$$

where $c(n)$ is the output to the motor at time n, $c(n)$ the error at time n, $e(n-1)$ the error at the previous sample, $c(n-1)$ the previous sample time output and k, A, and B are digital filter coefficients generally assumed positive.

This control equation shows that as repositioning begins, the position error and output value can be relatively large because $e(n-1)$ and $c(n-1)$ are zero. As position error decreases, the previous error and previous output values dominate. The equation becomes negative, and the output voltage becomes negative, which slows the motor. The constants A and B determine when braking occurs, that is, when the right hand side of the equation goes to zero.

The control equation we are discussing here has a transfer function form that follows from z-transforming both sides of it and collecting terms:

$$\frac{C(z)}{E(z)} = \frac{z - kA}{z + B}$$

We present in Fig. 1.7 a picture of a circuit board on which is implemented a microcontroller used to control a variety of different motor types. Essentially, these microcontrollers, among other things, implement equations such as the control equation in Example 1.2, although real controller equations are generally of a more complex form.

Figure 1.7   Multi-axis motion controller board that controls dc brush, dc brushless, ac induction, stepper, and variable-reluctance motors. The board is based on the Motorola DSP56001 (Courtesy of Ziatech Corporation.) See Reference [6].

1.3 CLASSICAL CONTROL VERSUS MODERN CONTROL

The digital and analog control systems represented in Figs. 1.1 and 1.2 and illustrated in Examples 1.1 and 1.2 are usually expressed in terms of frequency-domain transfer functions which are ratios of Laplace transforms or z-transforms (i.e., functions of the frequency variables s or z). Given a transfer function, the differential or difference equation relating input and output is directly available as well. Design and analysis of such systems employing, for example, the popular root-locus, Nyquist, and Bode techniques is termed *classical control theory*. Most of these methods have existed for more than 50 years. World War II was no doubt the main stimulus for the development and consolidation of classical control theory into a cohesive set of analysis and design methodologies. In particular, the servo control of antiaircraft weaponry required unprecedented technical sophistication, which was manifest in novel analysis and design techniques. And what was novel 50 ago we now call classical control theory.

We focus on the root-locus and Bode techniques for control system design. Although our concern is with digital control systems, most of the development of classical control theory has centered on analog systems. However, the discrete versions of the root-locus and Bode methodologies are almost identical to the analog versions. The rules of root-locus construction are identical for discrete and analog systems except that one is done in the z-plane and the other in the s-plane. Unlike analog Bode plots, which require an infinite frequency variation, discrete Bode plots call for a discrete frequency that is varied only over a finite range.

On the other hand, instead of using frequency-domain approaches for control system analysis and design, we can represent digital and analog control systems in the time domain by employing state-variable techniques and state-space models. Internal as well as external states of a system are revealed in such models. Only external inputs and outputs are generally available in classical control models. State-variable methods are taken up in detail in Chapter 3. These methods became popular in the 1960s and are included in what is now called *modern control theory*. The aerospace boom during the 1960s provided the impetus for the lively growth and development of modern control theory. The aerospace boom was stimulated by the American response to the Soviet Union's success in launching Sputnik in the late 1950s. President Kennedy's pledge in the early 1960s to land a human on the moon before the end of the decade guaranteed tremendous amounts of federal funding to aerospace companies. But even after the moon landing and the cutting back of federal funds, modern control theory continued to prosper. New techniques and methodologies appeared continually in the controls literature and many were applied to industrial control problems and have become part of the standard arsenal of practicing controls engineers. Almost all of these modern control techniques employ state-space ideas.

State-variable models of control systems lend themselves nicely to computer solutions. In fact, the development of state-variable theory for design and analysis of control systems has paralleled the advance of digital computer technology. State-space approaches have certain advantages over frequency-domain approaches. For one thing, nonlinear systems can be modeled in state variable form. We consider some nonlinear issues in Chapter 3. Systems with several inputs and several outputs can be handled easily in state-space formats. In Chapter 6 we discuss some modern state-space techniques and methodologies in the context of digital controller design. State estimators involving observer theory, as well as the linear quadratic optimal control procedure, are among the modern control techniques discussed in Chapter 6. Both classical and modern control techniques find extensive application in the practice of contemporary control systems engineering. We discuss both sets of techniques throughout this book.

One very novel and promising control strategy involving *fuzzy logic controllers*, which can be employed in both state-variable and classical frameworks, is considered briefly in Chapter 6. Fuzzy logic controllers have been popular in Japan for several years and are now beginning to be employed more and more by American engineers. The fuzziness of this procedure mirrors the fact that we typically lack complete information about reality. The "crispness" of reality often eludes us. The fact that there now exists a conceptual apparatus that permits us to deal with this lack of crispness is very exciting, especially because fuzzy logic controllers have been shown to be effective in controlling a wide variety of systems in a stable and robust fashion.

Example 1.3 Digital Filter

Fig. 1.8 represents a *second-order recursive* or *IIR* digital filter which we can cast into the state-variable form. The other major category of digital filters is called *nonrecursive* or *FIR*. These filters, however, are seldom used in closed-loop control systems because of the excessive delay associated with them.

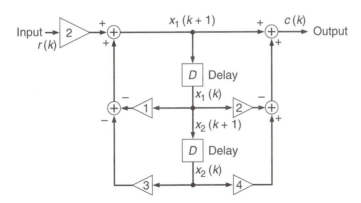

Figure 1.8   Block diagram of a digital filter.

We can write the equations that describe the system by looking at the outputs of the summing devices. Although we discuss the elements of such block diagrams in detail in Chapter 2 it should be fairly obvious that at the middle-left summer we have at the output $-x_1(k) - 3x_2(k)$ and at the middle-right summer we have $-2x_1(k) + 4x_2(k)$. Then at the top left we have $2r(k) - x_1(k) - 3x_2(k)$ which is $x_1(k+1)$. At the top right we have $c(k)$ which is $x_1(k+1) + (-2x_1(k) + 4x_2(k))$. Also from the middle of the block diagram we have $x_2(k+1) = x_1(k)$. We can rearrange these equations into the form

$$x_1(k+1) = -x_1(k) - 3x_2(k) + 2r(k)$$

$$x_2(k+1) = x_1(k)$$

and

$$c(k) = -3x_1(k) + x_2(k) + 2r(k)$$

and furthermore, we can put the equations into vector / matrix form as follows:

$$\begin{bmatrix} x_1(k+1) \\ x_2(k+1) \end{bmatrix} = \begin{bmatrix} -1 & -3 \\ 1 & 0 \end{bmatrix} \begin{bmatrix} x_1(k) \\ x_2(k) \end{bmatrix} + \begin{bmatrix} 2 \\ 0 \end{bmatrix} r(k)$$

and

$$c(k) = \begin{bmatrix} -3 & 1 \end{bmatrix} \begin{bmatrix} x_1(k) \\ x_2(k) \end{bmatrix} + [2]r(k)$$

These equations are the state-variable equation and the output equation in standard form, about which we have much more to say in Chapter 3.

1.4 DESIGN PROCESS OVERVIEW

Design is the process at the heart of engineering practice. The design of a digital control system typically consists of several steps starting from and returning to the realm or world of our everyday concerns and involvements. An idea emerging from this world drives a multidimensional and multistage process (the engineering design process), which has feedbacks between and among all stages. The actual contents of the design process can be unfolded in several ways. For example, according to W. E. Wilson:

"Among the most important features of the design process are the formulation of a mathematical model, the analysis of the sensitivity of the system with respect to its elements, the analysis of the compatibility of the various components and subsystems, the determination of the stability of the system when subjected to various inputs, optimization of the design with respect to some preselected criterion, prediction of the performance of the system, and the evaluation and testing of the system by means of a mathematical model or prototype [2]".

The contents of the design process are always grounded in a context, a world of some kind or other [3]. The environment, for instance, should always be part of a designer's concern. We will indicate our understanding of the multistaged design process as shown Fig. 1.9. The following stages are included:

- *The World.* A need or desire to design a control system does not just appear out of nowhere. The human context includes the social, political, economic, and environmental dimensions of human concern. Out of this broad-based context of involvement comes the general design problem to be solved, as well as specific requirements that focus a design.

- *System specification.* Design parameters specify control system designs. These parameters determine dynamic response, steady-state accuracy, and system stability. In both continuous and discrete domains, designs are usually specified by parameters such as peak time, percent overshoot, settling time, and rise time. (In Chapter 4 we cover these in detail.)

- *Mathematical modeling of the physical system.* The goal in modeling of a physical system is to convert the description of a system into a mathematical model. The most widely used model with classical control methods is a transfer function model. However, the transfer function model of a system describes only the system's input–output relationships. A more complete model is the state-space model of a system. This model provides the input–output relationships, but also the internal description of a system. In Chapter 3 we discuss the state-space model. CAE tools are recommended for modeling of physical systems especially as the order of the system exceeds 2 or 3.

- *Analysis.* Once it is available, a model of a system can be investigated from several points of view. Analysis of the mathematics representing the system reveals important information about the system. In Chapter 3 we cover the analysis of discrete control systems. This includes stability and sensitivity analysis of control systems.

- *Detailed Design.* In Chapters 4 and 6 we discuss the detailed design of digital control systems, and specifically, digital controllers using both conventional classical control methods (Chapter 4) and modern techniques (Chapter 6). Some of the more recent techniques such as fuzzy logic control and state-estimator design are investigated.

- *Verification.* CAE simulation packages are available for verification of digital control systems. For the design and analysis of discrete systems, we focus on the MATLAB and MATRIX$_\text{X}$ CAE packages. One very useful feature of both packages is the HELP statement, which reveals in a user-friendly fashion the details of any procedure. For example, typing HELP BODE will show the user how to enter numerators and denominators of transfer functions whose Bode plots are of interest.

- *Implementation.* The implementation of DSPs as single-chip digital controllers is discussed in Chapter 5. It should be mentioned that if we have a simple controller to implement, it might be easier to use PLDs (programmable logic devices) instead of the more complex DSP device. A PLD is a circuit that can be set up to perform a logic function, and if we have only a few logic functions to deal with, a PLD-based solution may be preferred. DSPs are generally preferred if we need to implement a complex algorithm requiring extensive and repetitive multiplication and addition operations. Additionally, PLDs may occasionally support DSP controllers by performing some of the functions common to DSPs. These include bus control and address decoding.

- *The World.* After the implemented design is tested, manufactured, and marketed, it enters the world as a product that is never without its impact on the world. Part of the designer's concern is to see that his or her products contribute to the good of the world. Such concern brings the design process into contact with the always important but often neglected field of engineering ethics.

The trial-and-error part of the design process has become much less tedious with the advent of CAE tools. CAE tools allow us to complete multiple design iterations. For example, if design flaws appear during the verification process, CAE tools allow us readily to modify the design parameters and verify the design again.

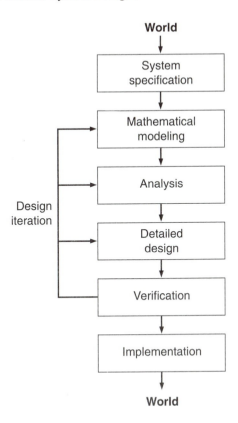

Figure 1.9   Design process for a digital control system.

1.5 ROLE OF DSPS IN CONTROL SYSTEM DESIGNS

As Fig. 1.2 indicates, a DSP can be used as a digital controller. This could be either a single-chip DSP or a system containing several DSPs. A DSP basically replaces analog controllers and conventional microprocessors in digital control system applications. In Chapter 5 we compare both analog and digital signal processors and DSPs and general-purpose microprocessors (see Tables 5.2 and 5.3).

Although analog signal processing has been around for a long time and is generally more familiar to designers, it is being used less in today's signal-processing applications. Because DSPs use software to control signal processing, they are more versatile than analog signal processors. For instance, once a set of design equations is determined and implemented on a DSP, the equations can easily be modified by changing a line or two of software code. This is not the case with analog controllers, which would usually require changing a hardware component, such as a capacitor or resistor.

However, we should keep in mind the PLD alternative to DSPs. If a controller can be realized with just a few logic functions, a DSP solution, which is expensive and complex to program, may be unnecessary. As mentioned in Section 1.4, a PLD is an integrated-circuit (IC) chip that can be configured and therefore customized by the user to implement a logic function. This function is a Boolean expression that contains registered and/ or combinational logic. PLDs offer very effective solutions in speed and density and are widely used in digital electronics over conventional standard TTL (transistor-transistor-logic) devices. A typical architecture of a PLD is shown in Fig. 1.10, where inputs are fed into an AND array and the resulting product terms are input to an OR array. In a PLD, the AND array is normally programmable and the OR array is fixed. Two examples of types of PLDs are the 16R8 and the 22V10.

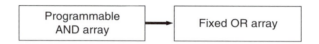

Figure 1.10   Architecture of a typical PLD.

In control system designs, PLDs can be used in the design of analog-to-digital converters, shaft encoders, and stepper motors. Because individual state machines can be defined easily on PLDs using PLD languages such as ABEL and CUPL, PLDs can be used as stepper motor controllers and shaft encoders. In designs where single-chip DSPs are employed, PLDs can be used to support the DSPs by performing address decoding, interrupt control, and bus control.

Complex PLDs (CPLDs) are like PLDs because they are user-configurable ICs, but they have more logic capabilities and complex functions. Fig. 1.11 shows the architecture of a typical CPLD; the architecture consists of a programmable interconnect structure that connects series of macrocells (building blocks of logic). The macrocells consist of an AND array followed by an OR array and a programmable register device such as a flip-flop. Fig. 1.12 shows a macrocell in a CPLD. Examples of CPLDs are devices from Xilinx, Inc. and Altera Corporation.

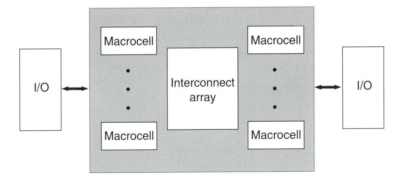

Figure 1.11   Architecture of a CPLD.

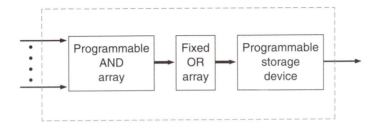

Figure 1.12   a Macrocell in a CPLD.

Designers normally regard PLD or CPLD solutions as special cases of DSP solutions. What a PLD does could certainly be done by a single-chip DSP, but we should make sure that the added cost and programming complexity of a DSP solution are necessary. Nevertheless, with the ever-growing complexity that the systems designers are called upon to control, the DSP approach is gaining favor all around the world. In addition, applications using DSPs that are supported by PLDs in a building-block fashion can add even more functionality to a single-chip DSP solution. We summarize some of the benefits of DSP and DSP/PLD solutions compared with traditional analog controllers:

- The control strategy can be changed easily without new hardware.
- In many control systems, undesired mechanical behaviors can be cancelled or compensated for by executing software.
- Undesired performance of a controller can be canceled or compensated for to improve total system performance.
- The system is smaller, lighter, requires less power, and costs less.
- The system is more reliable, maintainable, and testable.
- The system is more immune to noise.
- Complex control algorithms for higher control performance can be made.

Example 1.4 Active Vehicle Suspension System

DSPs can be used to control vehicle suspension systems. Fig. 1.13 shows a vehicle suspension system controller.

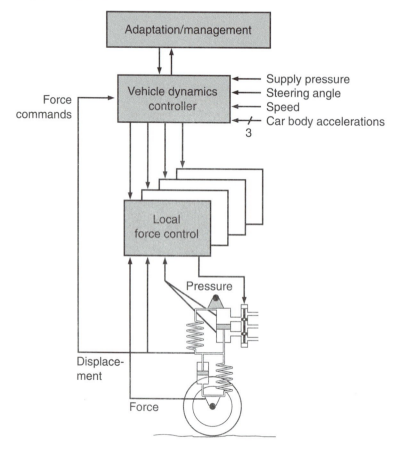

Figure 1.13   Active suspension control system (Courtesy of dSPACE GmbH.) See Reference [4].

This type of control system usually involves several sensors, four hydraulic actuators, and many mechanical degrees of freedom of the vehicle. It usually involves sampling rates in the kilohertz range for the hydraulic subsystems and controllers of order of 20 to 50 that have nearly 1000 coefficients.

Conventional analog controllers for such systems consist of passive spring damper mechanisms. However, high-order multivariable suspension controllers can be implemented using DSPs. The TMS32010, TMS32020, and TMS320C25 chips have been used as DSP controllers for this purpose. These are the fixed-point DSP chips which are relatively inexpensive but they do require that the designer scale the control algorithm. There are also floating-point DSP chips that can be used in these applications. They are more expensive but they do eliminate the need to scale the control algorithm, eliminating overflow and round-off problems as well. DSPs are also used as controllers for other automotive applications such as antiskid breaking, engine control, transmission control, and adaptive ride control.

Example 1.5 Disk-Drive Servo-Control

Fig. 1.14 shows the block diagram of a servo-control system for a disk-drive system where the servomotor controls the position of the read/write head on the disk.

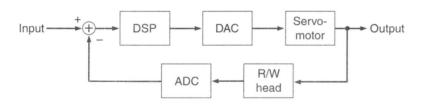

Figure 1.14   Disk-drive servo-control.

High sampling rates and complex control algorithms in such systems call for DSP solutions. A detailed example of a disk-drive servo-control system is presented in Example 5.3.

Example 1.6 Compliant Articulated Robot Vibration-Damping and Tracking Control [2]

A multivariable controller for a compliant articulated robot has been implemented using tools developed by the German company dSPACE. These tools are discussed in detail in Appendix B. The DSP robot controller for such a system needs to provide steady-state accuracy and good damping of structural vibrations at a high sampling rate, typically in the kHz range. A sketch of this system controlled by dSPACE DSP hardware hosted in a

---

[2] This example is presented by courtesy of dSPACE GmbH. See Reference [4].

standard PC/AT environment is shown in Fig. 1.15. This problem is discussed in more detail in Example 5.2.

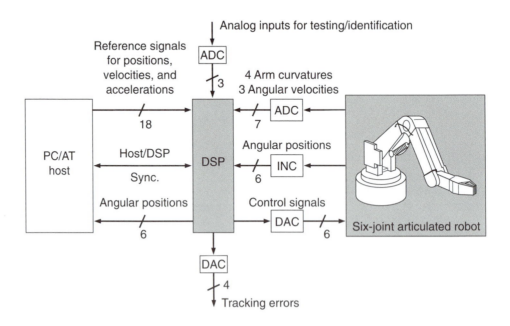

Figure 1.15   Robot vibration-damping and tracking control system.

Example 1.7 CPLDs Improve Motor Position Control System Design

We close this section with an example of how PLDs might be used in motor control. Complex PLDs (CPLDs) can replace parts of a closed-loop motion control system design. CPLDs offer flexibility, improve system performance and cost-effectiveness, and shorten development time compared to conventional circuit implementations.

A basic motor position control system consists of four main functions: signal converters (A/D and D/A), a microcontroller, a motor-load unit, and a feedback circuit. Fig. 1.16 shows a basic motor position control system. The A/D and D/A functions convert analog-to-digital and digital-to-analog signals where required within the control system. The microcontroller manages the dynamic behavior of the motor-load unit and stabilizes the system based on information from the input signal and the feedback path. The microcontroller is usually a single-chip DSP. However, for high-speed low-volume applications, an alternative to a DSP is a building-block method. This method uses multiple components to implement basic DSP functions, such as multipliers, accumulators, and shift and ALU operations.

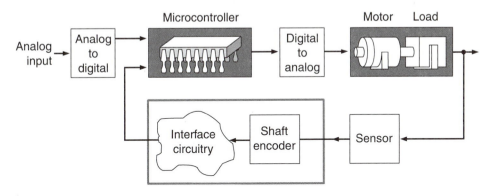

Figure 1.16   Basic motor position control system.

The feedback path components include a sensor, a shaft encoder, and interface circuitry. The sensor measures the motor-load shaft position using optical techniques and produces digital pulse signals as input to the shaft encoder. The shaft encoder transforms these signals into usable information for the microcontroller through the interface circuitry.

Shaft encoders measure the angle of rotation by counting clock pulses from a sensor. In the case of a synchronous, two-channel shaft encoder, which is shown in Fig. 1.17, the encoder circuitry measures the phase relationship between inputs A and B. When A leads B by 90 degrees, the encoder generates an up output. When A lags B by 90 degrees, the encoder generates a down output. Fig. 1.18 shows the input phase relationship. The up and down outputs represent clockwise and counterclockwise directions of rotation, respectively. The value of the rotational position is found by accumulating the output pulses of the encoder using an up/down counter.

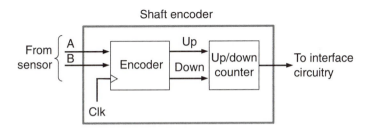

Figure 1.17   Synchronous two-channel shaft encoder.

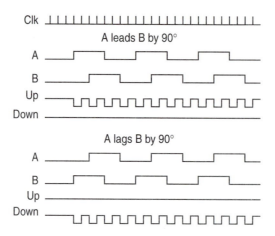

Figure 1.18   Input phase relationship.

We can implement the shaft encoder portion of the design shown in Fig. 1.16 using simple PALs (programmable array logic devices) and TTL components. However, this approach requires several devices; for example, three 16R8 PALs and two up/down counters are needed to implement just the shaft encoder. The interface circuitry requires additional logic.

Using CPLDs, we can replace the entire shaft encoder and interface circuitry portion of the feedback path with a single CPLD device, as illustrated in Fig. 1.19. Furthermore, if we use the building-block approach to implement the microcontroller, a CPLD device has even more advantages. In addition to incorporating the shaft encoder and interface circuitry, we can also perform various DSP functions, such as address decoding, DMA control, bus interface, interrupt control, and basic handshaking between building blocks.

Some possible solutions are the XC7200 and XC7300 family of devices from Xilinx or the MAX5000 and MAX7000 series from Altera.

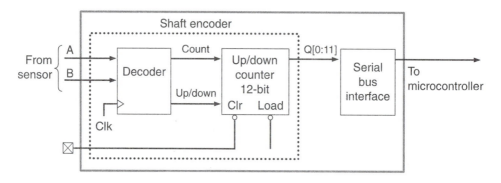

Figure 1.19   A CPLD replaces shaft encoder and interface circuitry.

1.6 CAE TOOLS

Before the widespread employment of digital computers, the design and analysis of control systems was a tedious business at best. Today all that has changed. Several CAE tools are presently available for the analysis and design of control systems. These tools allow us to perform complex computations and generate detailed plots with very little effort. Iterative design procedures especially are expedited. Nevertheless, we believe that the basic steps in procedures such as the root-locus should be able to be performed without a computer for low-order systems before embarking on high-order system computer-assisted solutions. Otherwise, depth of understanding is lost and interpretation of complex CAE solutions becomes troublesome. Still, the computer is a powerful ally. In fact, we can use CAE and CACE (computer-aided control engineering) tools for the following:

- Modeling of physical systems
- Stability analysis in the continuous domain
- Stability analysis in the discrete domain
- Conventional design techniques, such as root-locus, Routh Hurwitz, and Bode plots
- Modern design techniques, such as pole placement, controllability, and observability
- Transfer function to state-space form transformation, and vice versa
- s-to-z mapping, and vice versa
- Partial-fraction (residue) calculations
- Laplace transformation
- z-transformation
- Fourier and fast Fourier transforms
- Implementation of DSPs

In Appendices A and B we describe the CAE tools MATRIX$_X$ (which contains control design and signal processing modules), MATLAB (which contains control system and fuzzy logic toolboxes), and dSPACE (DSP-CITpro/eco software and hardware). In addition to the above, development tool sets from DSP vendors are available for the development and implementation of DSPs as microcontrollers. (Chapter 5 covers these tools.)

1.7   SUMMARY

In this chapter we presented an overview of digital control system using DSP. We presented basic differences between digital and analog control systems, as well as classical and modern control systems analysis and design techniques. Several qualitative examples were presented, among which were a robot control system, an active suspension control system, a disk-drive servo control system, and a motor position control system using PLDs. The intention was to engage the reader's interest so that she or he would be inspired enough to pursue further the issues at hand. The role of DSPs and different steps in the design process were discussed. We also discussed CAE design packages available for analyzing and designing digital control systems.

REFERENCES

1. A. Borgmann, "Information and Reality at the Turn of the Century," *Design Issues*, vol. XI, no. 2, Summer 1995, pp. 21–30.
2. W. E. Wilson, *Concepts of Engineering Systems Design,* New York, McGraw-Hill, 1965, p. 5.
3. G. Moriarty, "Engineering Design: Content and Context," *Journal of Engineering Education,* vol. 83, no. 2, April 1994.
4. *dSPACE Product Information,* dSPACE GmbH, Paderborn, Germany, 1990.
5. *MATRIX$_X$/System Build Tour Guide*, Integrated Systems, Inc., Santa Clara, CA, 1990.
6. *Technical Data Book,* Ziatech Corporation, 1991.
7. *Digital Control Applications with the TMS320 Family: Selected Application Notes,* Applications Book, Texas Instruments, Dallas, TX, 1991.
8. D. Yasuhiko, *Servo Motor and Motion Control Using Digital Signal Processors,* Prentice Hall, Upper Saddle River, NJ, 1990.
9. *The Programmable Logic Data Book,* Xilinx, Inc., San Jose, CA 1993.

Mathematical Methods of Discrete Systems

2.1 INTRODUCTION

The mathematical methods used to analyze discrete and continuous systems are similar. A set of linear differential equations describes the dynamics of a linear continuous-time system. A Laplace transform simplifies the set into algebraic equations, then inverse Laplace transforms are used to recover the signals. A set of difference equations describes the dynamics of a linear discrete-time system. A z-transform simplifies the set into algebraic equations, then inverse z-transforms are used to recover the signals.

A discrete-time system is defined as a system whose inputs and outputs are defined only at discrete instants of time. A discrete signal is typically generated from a continuous signal using a continuous-to-discrete device, an analog-to-digital converter, which requires that a continuous signal be sampled before being converted to discrete form. The most widely used sampling process is a uniform sampling in which the sampling switch closes every T seconds. Fig. 2.1 shows a simple continuous-to-discrete device.

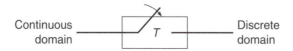

Figure 2.1   Continuous-to-discrete device.

In Chapter 3 we discuss ADC as well as DAC (digital-to-analog) devices.   An industrial ADC is shown in Fig. 2.2. In Section 2.2 of this chapter we consider difference equations and their block diagram representation. Section 2.3 covers the unit pulse response and discrete convolution.

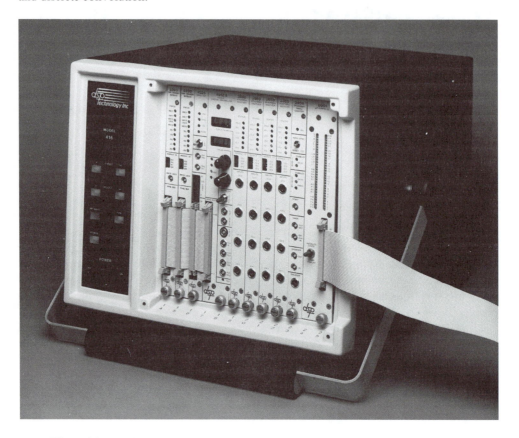

Figure 2.2   Industrial ADC (Courtesy of DSP Technology, Inc.) See Reference [12].

In Section 2.4 we present the z-transform (which transforms a difference equation into a simple algebraic equation), z-transform properties, and inverse z-transforms. Readers may want to refer to Appendix C (a table of z-transforms) and Appendix D (the partial-fraction expansion method) in conjunction with this section.

In Section 2.5 we discuss the discrete system transfer function representation of linear discrete systems, and in Section 2.6 we examine frequency response methods for systems and Fourier transforms, especially the discrete Fourier transform (DFT), for signals. In Section 2.7 we investigate the relationship between the s-plane and the z-plane and show various ways to map from the s-plane to the z-plane.

2.2 DIFFERENCE EQUATIONS

Linear difference equations are to discrete systems as differential equations are to continuous systems. In this section we discuss the classical method used to solve linear difference equations. A block diagram of a very simple linear discrete system is presented in Fig. 2.3.

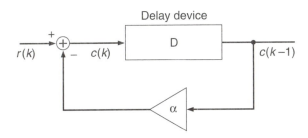

Figure 2.3   Simple block diagram.

This block diagram not only implements a simple difference equation but also illustrates the three basic elements used in all block-diagram representations of even the most complex linear discrete-time system. In terms of inputs $x(k)$ and outputs $y(k)$ these elements are:

- *Delay.* A delay element is a device that generates an output $y(k) = x(k-1)$ from the input $x(k)$. The output is delayed by one unit of discrete time. The delay device is often represented by a square block with a capital D inside (or with a z^{-1} inside, for reasons we will see when we study the z-transform.)

- *Multiplier.* A multiplier element is a device that generates an output $y(k) = \alpha x(k)$ from the input $x(k)$. The output is equal to the input multiplied by a constant α. The symbol for a multiplier is usually a triangle or a circle.

- *Summation.* A summation element is a device that generates an output $y(k) = x_1(k) + x_2(k)$ from the inputs $x_1(k)$ and $x_2(k)$. In general, the output is the summation or negative summation of n inputs.

From Fig. 2.3, at the output of the summation device, we can write

$$c(k) = r(k) - \alpha c(k - 1) \qquad (2.1)$$

$$c(k) = \alpha c(k - 1) = r(k) \qquad (2.2)$$

This is a first-order difference equation because in addition to $c(k)$ there is a term with a shift of one unit of discrete time away from $c(k)$. A first-order difference equation with one shift is analogous to a first-order differential equation with one derivative term. Like a differential equation, we can solve a linear difference equation by distinguishing a homogeneous solution $c_h(k)$ and a particular solution $c_p(k)$ where

$$c(k) = c_h(k) + c_p(k) \qquad (2.3)$$

Let us simplify our equation by particularizing it and letting $\alpha = -0.25$ and $r = 0.5^k$ and assuming c(0)=1 as our initial condition. As in the solution of differential equations, we need n initial conditions for an nth-order difference equation. The difference is that the differential equation has initial conditions as derivative terms evaluated at say t=0 while a difference equation has initial conditions as shifted terms, such as c(0), c(-1), and c(-2) for a third-order difference equation.

To solve for $c_h(k)$ we set the right-hand side (RHS) of the difference equation to zero and let $c_h(k) = \alpha_1(r)^k$. Then

$$c(k) + \alpha c(k - 1) = 0 \qquad (2.4)$$

becomes

$$\alpha_1(r)^k + \alpha \alpha_1(r)^{k-1} = 0 \qquad (2.5)$$

from which

$$r + \alpha = 0 \qquad\qquad (2.6)$$

emerges as the characteristic equation. Then

$$r = -\alpha = +0.25 \qquad\qquad (2.7)$$

is the root of the characteristic equation and $c_h(k)$ takes the form

$$c_h(k) = \alpha_1(0.25)^k \qquad\qquad (2.8)$$

Now, to solve for $c_p(k)$, we look at the RHS of Equation 2.2 and let $c_p(k)$ be of the form of the RHS plus the form of all possible shifts of the RHS. Again, this is like the differential equation where we look at the RHS and all the possible derivatives of the RHS to get the form $c_p(t)$.

Here we have $A(0.5)^k$ and all possible shifts of this form yield the same basic form. Therefore, we let

$$c_p(k) = A(0.5)^k \qquad\qquad (2.9)$$

Then

$$c_p(k-1) = A(0.5^{k-1}) \qquad\qquad (2.10)$$

and

$$A(0.5^k) + (-0.25)A(0.5)^{k-1} = 0.5^k \qquad\qquad (2.11)$$

from which

$$A = 2 \qquad\qquad (2.12)$$

Then

$$c_p(k) + c_h(k) = 2(0.5)^k + \alpha_1(0.25)^k = c(k) \qquad\qquad (2.13)$$

Evaluated at $k = 0$ we get

$$c(0) = 2 + \alpha_1 = 1 \tag{2.14}$$

so

$$\alpha_1 = -1 \tag{2.15}$$

Then

$$c(k) = 2(0.5)^k - (0.25)^k \tag{2.16}$$

is the total solution of the difference equation, which is actually valid for all integers k from $-\infty < k < \infty$.

We can generalize these results by proposing a general form of an nth order difference equation as

$$c(k) + \sum_{i=1}^{n} \alpha_i c(k-i) = \sum_{j=0}^{m} \beta_j r(k-j) \tag{2.17}$$

for $k = (0,1, 2, 3, \ldots)$. The constants α_i and β_i come from the discrete system that Equation 2.17 represents. Fig. 2.4 summarizes the approach used to find a solution to a linear difference equation.

Example 2.1

Find the solution for the following difference equation:

$$c(k) - 16c(k + 1) + 15c(k - 2) = 0$$

with

$$c(-1) = 2 \text{ and } c(-2) = 2$$

Solution : The characteristic equation is

$$r^2 - 16r + 15 = 0$$

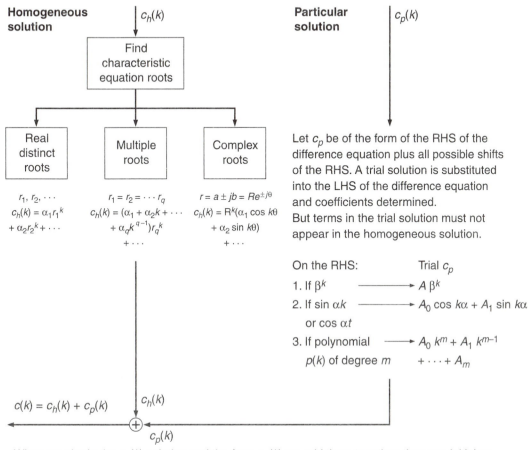

Figure 2.4   Approach used to find a solution to a linear difference equation.

and the roots of this equation are r = 1 and r = 15. Therefore, the homogeneous solution, the total solution in this case, is

$$c(k) = \alpha_1 + \alpha_2 15^k$$

and using the initial conditions, we get

$$c(-1) = 2 = \alpha_1 + \alpha_2 (15)^{-1} \text{ and } c(-2) = -1 = \alpha_1 + \alpha_2 (15)^{-2}$$

from which $\alpha_1 = -1.21$ and $\alpha_2 = 48.21$. Thus

$$c(k) = -1.21 + 48.21(15)^k$$

Example 2.2

Find the solution for the following difference equation:

$$c(k) + 2c(k-1) + 2c(k-2) = 0$$

with $c(-1) = c(-2) = 1$

Solution : The roots of the characteristic equation are complex:

$$r = \sqrt{2}e^{-j\frac{3\pi}{4}} \text{ and } \sqrt{2}e^{+j\frac{3\pi}{4}}$$

Therefore, we let

$$c(k) = 2^{\frac{k}{2}}(\alpha_1 \cos\frac{3\pi}{4}k + \alpha_2 \sin\frac{3\pi}{4}k)$$

and

$$c(-1) = \frac{1}{\sqrt{2}}(\alpha_1(-\frac{1}{\sqrt{2}}) - \alpha_2(\frac{1}{\sqrt{2}})) = 1$$

$$c(-2) = \frac{1}{2}(\alpha_1(0) - \alpha_2(-1)) = 1$$

from which $\alpha_1 = -4.0$, $\alpha_2 = 2.0$ and

$$c(k) = 2^{\frac{k}{2}}(-4\cos\frac{3\pi}{4}k + 2\sin\frac{3\pi}{4}k)$$

Example 2.3

Find the solution for the following difference equation:

$$c(k) - 16c(k-1) + 15c(k-2) = 5k - 2$$

with

$$c(-1) = c(-2) = 0$$

Solution : The homogeneous solution for this difference equation, which we found in Example 2.1, is

$$c_h(k) = \alpha_1 + \alpha_2 15^k$$

To find the particular solution, we assume the trial solution $c_p(k) = A_0 k + A_1$
Substituting this solution into the difference equation yields

$$c(k) - 16c(k-1) + 15c(k-2) =$$

$$A_0 k + A_1 - 16(A_0(k-1) + A_1) + 15(A_0(k-2) + A_1) = 5k - 2$$

Collecting terms, we obtain

$$0(k) + (16 - 30)A_0 = 5k - 2$$

That is, $0 = 5$, which means that our trial solution failed. The reason for the failure is that the A_1 constant term is already included in the α_1 of the homogeneous solution. The next trial should increase the order of the polynomial:

$$c_p(k) = A_0 k^2 + A_1 k + A_2$$

Then we obtain

$$-28A_0 k + (44A_0 - 14A_1) = 5k - 2$$

from which $A_0 = -0.18$, $A_1 = -0.42$, A_2 can be anything, so let $A_2 = 0$. Then

$$c_p(k) = -0.18k^2 - 0.42k$$

and

$$c(k) = \alpha_1 + \alpha_2 15^k - 0.18k^2 - 0.42k$$

with α_1 and α_2 determined from the initial conditions. We get, finally,

$$c(k) = -0.11 - 1.93(15)^k - 0.18k^2 - 0.42k$$

A linear difference equation is the basic description of a linear discrete system. It is a very practical representation of the system and can easily be implemented using delays, multipliers, and summations.

Example 2.4

Use the basic three discrete system elements to represent the following difference equation with a block diagram:

$$c(k) = \beta_0 r(k) + \beta_1 r(k-1) - \alpha_1 c(k-1)$$

Solution: By noting the shifts in the input $r(k)$ and the output $c(k)$ on the right side of the difference equation above, we can implement its block diagram as in shown Fig. 2.5.

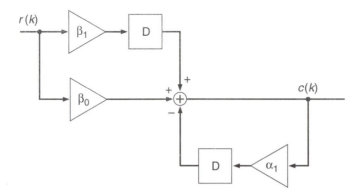

Figure 2.5   Block-diagram representation of the system in Example 2.4.

A summation element is used to sum the three signals $\beta_0 r(k)$, $\beta_1 r(k-1)$, and $\alpha_1 c(k-1)$. The signal $\beta_0 r(k)$ is generated by passing the input $r(k)$ through a multiplier. The signals $\beta_1 r(k-1)$ and $\alpha_1 c(k-1)$ are generated by passing them through delay and multiplier elements.

Another approach to difference equation solution is the iterative or recursive approach. A programmable calculator or computer can readily be employed in this type of solution. The basic idea is to set the difference equation into the form of Equation 2.17, then solve so that only $c(k)$ is on the LHS. Let $k=0$, and all the terms on the RHS are known. Then let $k=1$, and using the results from $k=0$ we again know all the terms on the RHS. Then let $k=2$, and so on. A problem with this approach is that we do not get a closed-form solution, but often only a numerical answer is needed.

Example 2.5

Find the solution of the following difference equation using iteration:

$$c(k) - 0.5c(k-1) = 0 \text{ with } c(-1) = 2$$

Solution:

$$c(k) = 0.5c(k-1)$$

$$c(0) = 0.5(2) = 1$$

$$c(1) = 0.5(1) = 0.5$$

$$c(2) = 0.5(0.3) = 0.25$$

$$c(3) = 0.5(0.25) = 0.125$$

and so on, with each value being half of the preceding value.

2.3 UNIT PULSE RESPONSE AND DISCRETE CONVOLUTION

The decomposition of $c(k)$ into a homogeneous solution and a particular solution is the classical mathematical approach to difference equation solution. Another more systems-oriented approach is also available. Instead of viewing $c(k)$ as just a solution to a difference equation, we can also view it as the response of a linear discrete-time system whose input is r(k). A block-diagram representation is shown in Fig. 2.6.

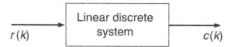

$r(k)$ → | Linear discrete system | → $c(k)$

Figure 2.6   Block diagram of a standard linear discrete system.

Here we decompose c (k) into a zero-input response $c_{zi}(k)$ and a zero-state response $c_{zs}(k)$. The c_{zi} term is obtained by setting r(k) = 0 for all k and solving for the response due to initial conditions. The c_{zs} term is obtained by setting the initial conditions to zero and solving for the response due to the input. Unlike the solutions for $c_h(k)$ and $c_p(k)$ in the classical approach, the c_{zi} and c_{zs} terms can be solved for independently.

Example 2.6

Find the solution of the following difference equation using zero-state and zero-input decomposition:

$$c(k) + 0.5c(k-1) = r(k)$$

with $c(-1) = 2$ and $r(k) = (-1)^k$.

Solution : When, $c_{zi}(k)$ takes a homogeneous form,

$$c_{zi}(k) = \alpha_1(-0.5)^k$$

then

$$c(-1) = 2 = \alpha_1(-0.5)^{-1} = -2\alpha_1$$

so $\alpha_1 = -1.0$ and $c_{zi}(k) = -(-0.5)^{-k}$. $C_{zs}(k)$ takes a complete solution form,

$$c_{zs}(k) = c_h + c_p$$

then

$$c_h = \alpha_2(-0.5)^k \text{ and } c_p = A(-1)^k$$

To get A:

$$A(-1)^k + 0.5A(-1)^{k-1} = (-1)^k$$

$A - 0.5A = 1$ or $A = 2$ and, $c_p = 2(-1)^k$ Then $c_{zs}(k) = \alpha_2(-0.5)^k + 2(-1)^k$
and with $c(-1) = 0$, we get $\alpha_2 = -1$. Therefore, $c_{zs}(k) = 2(-1)^k - (-0.5)^k$ and

$$c(k) = c_{zs} + c_{zi} = 2(-1)^k - 2(-0.5)^k$$

The c_{zi} solution is usually straightforward. The c_{zs} solution is generally as involved as the classical mathematical solution because both a homogeneous and a particular part need to be obtained. But there does exist an alternative to this method of c_{zs} solution. It involves discrete convolution: and convolution calls for a function called the *unit pulse response*, h(k), the response of a discrete with zero initial conditions to a *unit pulse function*. But what is a unit pulse function? It is sometimes called a *Kronecker delta sequence* and is defined below.

Kronecker Delta Sequence (Unit Pulse)

Probably the simplest function in the universe, a Kronecker delta sequence is defined as a sequence that has a value of 1 for k = 0 and has a value of 0 for all other integer units of the discrete system. Equation 2.18 and Fig. 2.7 define this sequence.

$$\delta(k) = \begin{cases} 1, k = 0 \\ 0, k \neq 0 \end{cases} \tag{2.18}$$

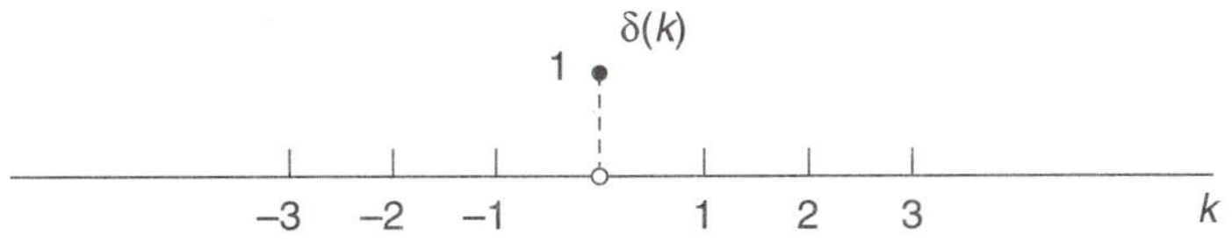

Figure 2.7   Kronecker delta sequence.

Another way to define a unit pulse function is in terms of its effect:

$$x(k) = \sum_{n=-\infty}^{\infty} x(n)\delta(k-n) \qquad (2.19)$$

The unit pulse samples $x(n)$ and produces the value at $n = k$. Instead of looking at Equation 2.19 as a definition of the Kronecker delta, we can interpret the equation another way. Equation 2.19 implies that every function $x(k)$ can be expressed as an infinite series of shifted unit pulses, each weighted with a constant term $x(n)$.

This is a good place to define two other discrete functions that play important roles in the design and analysis of many discrete systems: the unit step and unit ramp sequences.

Unit Step Sequence

A unit step sequence is defined by Equation 2.20 and is shown in Fig. 2.8.

$$u(k) = \begin{cases} 1, k \geq 0 \\ 0, k < 0 \end{cases} \qquad (2.20)$$

This sequence has a value of 1 for k greater than or equal to zero. It has a value of zero for k less than zero.

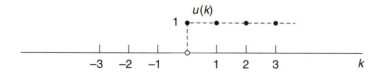

Figure 2.8   Unit step sequence.

Unit Ramp Sequence

A unit ramp sequence is defined by Equation 2.21. Fig. 2.9 shows this sequence.

$$r(k) = \begin{cases} k, k \geq 0 \\ 0, k < 0 \end{cases} \qquad (2.21)$$

This sequence increases linearly for positive discrete time and has a value of zero for negative discrete time.

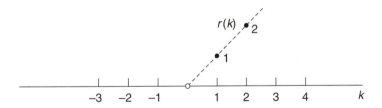

Figure 2.9   Unit ramp sequence.

Let us also define three other key concepts that will be relevant to our endeavors: *causality*, *superposition*, and *time invariance*.

- A discrete function or signal, say the output or input of a discrete system c(k) or r(k), is *causal* if it is zero for k < 0. A discrete system is causal if h(k) = 0 for k < 0, that is, if the system's unit pulse response is a causal function. Since h(k) is the response of a system to the input $\delta(k)$, causality means that we can have no output prior to the application of the input. A noncausal system is sometimes called *anticipatory* because it has some response prior to application of the input. We will normally assume causality. Causal systems are physically realizable and can be implemented with standard discrete system hardware: for example, summers, delays, and multipliers.

- *Superposition* incorporates the two concepts of homogeneity and additivity. The former concept implies that $\alpha r(k)$ in Fig. 2.6 yields $\alpha c(k)$, where α is a constant. The latter concept implies that $r_1(k) + r_2(k)$ at the input yields $c_1(k) + c_2(k)$ at the output, as long as $r_1(k)$ yields $c_1(k)$ and $r_2(k)$ yields $c_2(k)$. To say that a system satisfies superposition is to say that it is linear. When $r_1(k)$ yields $c_1(k)$ and $r_2(k)$ yields $c_2(k)$, then $\alpha r_1(k) + \beta r_2(k)$ yields $\alpha c_1(k) + \beta c_2(k)$ for any r_1 and r_2 with α and β constants. Most of the systems we deal with will be assumed linear. If a system is not linear, it is said to be *nonlinear*, and many of these kinds of systems can be approximated by linear models via a process of linearization. We consider the linearization process in Chapter 3 after we introduce the state variable format for dynamic systems.

- We will normally assume that our systems are time invariant. *Time invariance* means that the structure of the system does not change over time. In the difference equation model, the coefficients that reflect the system structure will remain fixed constants. Also implied by this is that if r(k) yields c(k), then r(k-n) yields c(k-n); that is, the output retains its shape even if the input is delayed in time.

We now develop discrete convolution. Considering Fig. 2.6 again, $r(k) = \delta(k)$ implies that c(k) = h(k). Also, $\alpha\delta(k)$ implies that $\alpha h(k)$ for constant values of α, and $\alpha\delta(k-i)$ yields $\alpha h(k - i)$. Now let $\alpha = r(i)$ and sum both input and output over all i. Then $\sum_{i=-\infty}^{\infty} r(i)\delta(k-i)$ at the input yields $\sum_{i=-\infty}^{\infty} r(i)h(k-i)$ at the output. Since by a property of the unit pulse function the input reduces to r(k), the term for the general input to a discrete system, the output must become c(k), the general output of a discrete system, and we can write

$$c(k) = \sum_{i=-\infty}^{\infty} r(i)h(k-i) \tag{2.22}$$

which is the famous *discrete convolution summation*.

Let us assume that in Equation 2.22 our functions are causal, which is not a serious restriction for real systems because we need to start and stop our signals at some point in time and we can generally call the start time k = 0. Then Equation 2.22 can be written, after a minor change of variable, in several different but equivalent forms:

$$c(k) = \sum_{i=0}^{k} r(i)h(k-i)$$

$$c(k) = \sum_{i=0}^{k} h(i)r(k-i)$$

$$c(k) = h(k)*r(k) \tag{2.23}$$

$$c(k) = r(k)*h(k)$$

The last two expressions in Equation 2.23 are shorthand expressions for discrete convolution. Now, c(k) has been developed assuming zero initial conditions. The discrete convolution summation, then, gives us only c_{zs}. But that is often all we need.

Example 2.7

Find the zero-state response for the following difference equation using discrete convolution:

$c(k) + 0.25 \, c(k-1) = r(k)$, where $r(k) = (0.5)^k u(k)$

Solution: To get h(k), let $r(k) = \partial(k)$ and $c(k) = h(k)$:

$h(k) + 0.25h(k-1) = \delta(k)$

Since the RHS = 0 for k > 0, the h(k) solution might look like a homogeneous solution. Try $h(k) = A(-0.25)^k u(k)$ then

$$A(-0.25)^k u(k) + 0.25 A(-0.25)^{k-1} u(k-1) = \delta(k)$$

$$A(-0.25)^k [u(k) + 0.25(-0.25)^{-1} u(k-1)] = \delta(k)$$

$$A(-0.25)^k [u(k) - u(k-1)] = \delta(k)$$

But the term in brackets is $\delta(k)$, and

$$A(-0.25)^k \delta(k) = \delta(k)$$

$$A(-0.25)^0 \delta(k) = A\delta(k)$$

Thus A=1 and $h(k) = (-0.25)^k u(k)$. To get c_{zs}, use

$$c(k) = \sum_{i=0}^{k} r(i)h(k-i)$$

$$= \sum_{i=0}^{k} (0.5)^i u(i)(-0.25)^{k-i} u(k-i)$$

$$= (-0.25)^k \sum_{i=0}^{k} \left(\frac{-4}{2}\right)^i u(i)u(k-i)$$

The product of u(i)u(k-i) tells us to start at i = 0 and stop at i = k (which is already indicated by the summation) , but it also tells us that k must be greater than or equal to zero, which we incorporate by multiplying our result by u(k). Then using the formula

$$\sum_{i=0}^{k} \alpha^i = \frac{1-\alpha^{k+1}}{1-\alpha} \text{ , we get } \sum_{i=0}^{k}(-2)^i = \frac{1}{3} + \frac{2}{3}(-2)^k$$

Then $c_{ZS}(k)=[0.33(-0.25)^k+0.67(0.5)^k]u(k)$.

Example 2.8

Determine c_{zs} if $r(k) = (-1)^k u(k)$ and $h(k) = (-2)^k u(k-5)$.

Solution:

$$c_{zs}(k) = r(k) * h(k) = \sum_{i=0}^{k}(-2)^i u(i-5)(-1)^{k-i}u(k-i)$$

$$= (-1)^k[\sum_{i=5}^{k}(2)^i]u(k-5)$$

Here we can use the formula

$$\sum_{i=n_1}^{n_2} \alpha^i = \frac{\alpha^{n_1} - \alpha^{n_2+1}}{1-\alpha}$$

Then

$$c_{zs}(k) = (-1)^k(\frac{2^5 - 2^{k+1}}{1-2})u(k-5)$$

$$c_{zs}(k) = (-1)^k[2(2)^k - 32]u(k-5)$$

2.4 THE Z-TRANSFORM

With employment of the z-transform, the discrete convolution operation reduces to multiplication, and difference equation solution becomes a matter of algebra. In fact, the z-transform has applicability to many areas of discrete system analysis and design. In this section we define the z-transform, discuss essential theorems and properties, develop the inverse z-transform, and consider a few of the more important applications of the z-transform.

z-transforms are to discrete systems as Laplace transforms are to continuous systems. The z-transform of a signal f(k) is defined mathematically as

$$Z\{f(k)\} = F(z) = \sum_{k=-\infty}^{\infty} f(k)z^{-k} \qquad (2.24)$$

This is a double-sided z-transform and shows a power series in the variable z.

$$F(z) = \sum_{k=-\infty}^{-1} f(k)z^{-k} + \sum_{k=0}^{\infty} f(k)z^{-k} \qquad (2.25)$$

In Equation 2.25 we will assume that the first term on the right side of the equation is zero because the functions we want to transform will be causal. The equation reduces to a one-sided z-transform:

$$F(z) = \sum_{k=0}^{\infty} f(k)z^{-k} \qquad (2.26)$$

Equation 2.26 is the basic definition of the z-transform for the rest of this book. Noncausal functions are not usually needed in the study of deterministic systems. Probabilistic or stochastic systems studies do employ the two-sided z-transform to analyze noncausal signals such as auto- and cross-correlation functions. But such concerns are beyond our present scope.

Any signal that is Laplace transformable has a discrete version that is z-transformable. Thus the study of the z-transform and the inverse z-transform, which recovers f(k) from a given F(z), is analogous to the study of the Laplace transform and the inverse Laplace transform. Table 2.1 presents the z-transforms of the most common functions used in control systems.

Table 2.1   z-transforms of Common Functions Used in Control Systems

f(k)	F(z)
$\delta(k)$	1
$u(k)$	$\dfrac{z}{z-1}$
$k u(k)$	$\dfrac{z}{(z-1)^2}$
$a^k u(k)$	$\dfrac{z}{z-a}$
$\sin k\omega T\, u(k)$	$\dfrac{z\sin\omega T}{z^2 - 2z\cos\omega T + 1}$
$\cos k\omega T\, u(k)$	$\dfrac{z(z-\cos\omega T)}{z^2 - 2z\cos\omega T + 1}$

The u(k) multiplier can be suppressed if causality is assumed.

Example 2.9

The unit pulse function is

$$\delta(k) = \begin{cases} 1, k = 0 \\ 0, k \neq 0 \end{cases}$$

We can apply the z-transform definition in Equation 2.26 to get

$$F(z) = \sum_{k=0}^{\infty} \delta(k)z^{-k} = \sum_{k=0}^{\infty} \delta(k)z^0 = 1$$

Therefore, the z-transform of the unit pulse function is 1: $\delta(k) \leftrightarrow 1$. This double-arrow notation is a standard indication of a z-transform pair.

Example 2.10

The unit step function is

$$u(k) = \begin{cases} 1, k \geq 0 \\ 0, k < 0 \end{cases}$$

We apply Equation 2.26 to get

$$F(z) = \sum_{k=0}^{\infty} u(k)z^{-1} = \sum_{k=0}^{\infty} z^{-k} = \frac{z}{z-1}$$

This result uses the formula $\sum_{k=0}^{\infty} \alpha^k = \frac{1}{1-\alpha}$, $|\alpha| < 1$, thus

$$u(k) \leftrightarrow \frac{z}{z-1}$$

Example 2.11

The z-transform of $a^k u(k)$ is

$$F(z) = \sum_{k=0}^{\infty} a^k z^{-k} = \sum_{k=0}^{\infty} (\frac{a}{z})^k = \frac{1}{1-\frac{a}{z}} = \frac{z}{z-a}$$

Thus

$$a^k u(k) \leftrightarrow \frac{z}{(z-a)}$$

Now, the derivative of both sides of the expression in Example 2.11 yields additional z-transform pairs:

$$ka^{k-1} \leftrightarrow \frac{z}{(z-a)^2} \qquad (2.27)$$

Differentiating again yields

$$k(k\text{-}1)a^{k-2} \leftrightarrow \frac{2z}{(z-a)^3} \tag{2.28}$$

And again:

$$k(k\text{-}1)(k\text{-}2)a^{k-3} \leftrightarrow \frac{3!z}{(z-a)^4} \tag{2.29}$$

This process can be repeated to provide even more z-transform pairs. The next example illustrates some of these results.

Example 2.12

Find the z-transform of $ku(k)$ and $k^2 u(k)$.

Solution: The z-transform of $ku(k)$ is given as the third entry in Table 2.1. It follows directly from Equation 2.27 with $a = 1$. (Note that "Equations" 2.27 to 2.29 are really correspondences, not equations.)

$$k \leftrightarrow \frac{z}{(z-1)^2}$$

But the z-transform of $k^2 u(k)$ is not so obvious. We can write

$$k(k-1)a^{k-2} = (k^2 - k)a^{k-2} \leftrightarrow \frac{2z}{(z-a)^3}$$

and letting $a = 1$,

$$k^2 - k \leftrightarrow \frac{2z}{(z-1)^3}$$

Adding $k \leftrightarrow \dfrac{z}{(z-1)^2}$ to both sides of this expression, we get

$$k^2 \leftrightarrow \frac{2z}{(z-1)^3} + \frac{z}{(z-1)^2} = \frac{z^2 + z}{(z-1)^3}$$

Looking at the z-transform from Example 2.12, $(z^2 + z) / (z - 1)^3$ we note that when $z = 1$, $F(z)$ does not exist. In fact, since $F(z)$ is defined as an infinite series $\sum_{k=0}^{\infty} f(k)z^{-k}$ we must consider when such a series converges. The ratio test can normally be used to investigate convergence: If we have an infinite series $\sum_{k=0}^{\infty} a_k$, then if $\left|\dfrac{a_{k+1}}{a_k}\right| < 1$

as $k \rightarrow \infty$, we say that the series converges. For example, for $a^k u(k) \leftrightarrow \dfrac{z}{z - a}$, we can

write $a_k = a^k z^{-k} = (\dfrac{a}{z})^k$. Thus $\left|\dfrac{(\frac{a}{z})^{k+1}}{(\frac{a}{z})^k}\right| < 1$ implies $|a| < |z|$ and this is called the

region of convergence (ROC) for $F(z)$.

Like the Laplace transform $F(s)$, $F(z)$ is a function of a complex frequency variable. As the s-plane played an important role in continuous-time systems theory, the z-plane will be important in discrete-time systems theory. Since z is a complex variable, which we can represent in polar form as $z = re^{j\theta}$, $|z| = |a|$ represents a circle in the z-plane with radius $|a|$. Thus the ROC for $F(z) = z / (z - a)$ is the area outside that circle, as indicated in Fig. 2.10.

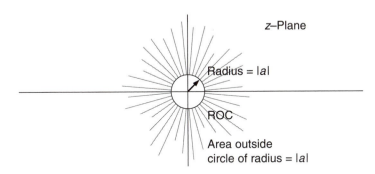

Figure 2.10 Indication of the ROC for $F(z) = z / (z - a)$.

For z equal to a/2, for instance, the infinite series that constitutes the z-transform does not converge and $F(z)$ does not exist. The ROC does play a crucial role in inverse z-transforms when the two-sided z-transform is involved and the time sequences are noncausal. But since causal functions are our major concern, we need not further pursue the theoretical issues involved in the notion of ROC.

Properties of the z-transform

In this section we present some of the important theorems or properties of the z-transform. Many of these are essential to z-transform applications and provide shortcut approaches that greatly simplify the work involved. Proofs of most of these theorems are left as exercises for the reader.

Theorem 2.1 (Multiplication Theorem). The z-transform of a sequence, multiplied by a real constant α, is equal to the constant multiplied by the z-transform of the sequence:

$$Z\{\alpha f(k)\} = \alpha Z\{f(k)\} = \alpha F(z) \tag{2.30}$$

Example 2.13

Find the z-transform of the sequence

$$f(k) = 10\cos k\omega T$$

Solution: The z-transform of the sequence above is

$$F(z) = Z\{10\cos k\omega T\} = 10 Z\{\cos k\omega T\} = 10\frac{z(z-\cos\omega T)}{z^2 - 2z\cos\omega T + 1}$$

Theorem 2.2 (Linearity Theorem). If F (k) and G (k) are two functions in the discrete domain, and F(z) and G(z) are the resulting z-transforms, then

$$Z\{\alpha f(k) + \beta g(k)\} = \alpha F(z) + \beta G(z) \tag{2.31}$$

where α and ß are constants.

Example 2.14

If $f(k) = 5\sin k\omega T + (k)$, we use the linearity theorem to obtain

$$F(z) = Z\{5\sin k\omega T\} + Z\{k\} = 5Z\{\sin k\omega T\} + Z\{k\}$$

$$= \frac{z \sin \omega T}{z^2 - 2z\cos \omega T + 1} + \frac{z}{(z-1)^2}$$

Theorem 2.3 (Convolution Theorem). Discrete convolution plays an important role in investigating the response of a discrete system to an arbitrary input signal. We saw that $c_{zs}(k) = r(k)*h(k)$. More generally, if $g(k) = f(k)*g(k)$, where $g(k) \leftrightarrow G(z)$, $f(k) \leftrightarrow F(z)$, and $q(k) \leftrightarrow Q(z)$, the z-transform of both sides of the equation yields

$$Z\{f(k)*g(k)\} = F(z)G(z) = Q(z) \tag{2.32}$$

Therefore, the z-transform of two convoluted sequences is the product of their corresponding z-transforms.

Example 2.15

If $f(k) = k^2$ and $g(k)$ comes from sampling $g(t) = e^{-2t}u(t)$ such that $t = kT$ and $g(k) = e^{-2kT}u(kT)$, find the z-transform of $f(k)*g(k)$.

Solution: Since $u(kT) = u(k)$,

$$g(k) = (e^{-2T})^k u(k) \leftrightarrow \frac{z}{z - e^{-2T}} = G(z)$$

and

$$k^2 \longleftrightarrow \frac{z^2 + z}{(z-1)^3} = F(z)$$

Therefore,

$$Z\{f(k)*g(k)\} = F(z)G(z) = Q(z) = \frac{z(z+1)}{(z-1)^3}\left[\frac{z}{z - e^{-2T}}\right] = \frac{z^2(z+1)}{(z-1)^3(z - e^{-2T})}$$

The problem with this result is that $Q(z) = F(z)G(z) \leftrightarrow q(k)$ and we need to get $q(k)$. That requires the inverse z-transform procedure, which will be considered shortly. The point is that even though the z-transform reduces convolution to multiplication, this simplification generally costs us the additional step of the inverse z-transform.

Theorem 2.4 (Initial Value Theorem). This theorem allows us to obtain the value of f(k) at k = 0 from a knowledge of F(z). From the definition of the z-transform,

$$F(z) = \sum_{k=0}^{\infty} f(k)z^{-k} \qquad (2.33)$$

we can expand the sequence above to obtain

$$F(z) = f(0) + f(1)z^{-1} + f(2)z^{-2} + \cdots \qquad (2.34)$$

The first term on the right side of Equation (2.34) is the initial value of the z-transformed function. Therefore,

$$f(0) = \lim_{z \to \infty} F(z) \qquad (2.35)$$

Example 2.16

Find the initial value of a sequence that has the z-transform

$$F(z) = \frac{z}{(z-1)^2}$$

Solution: We can use the initial value theorem to obtain

$$f(0) = \lim_{z \to \infty} F(z) = \lim_{z \to \infty} \frac{z}{(z-1)^2} = 0$$

Theorem 2.5 (Final Value Theorem). In many situations, we would like to know the final value of a sequence. The final value of a sequence f(k) is

$$f(\infty) = \lim_{z \to \infty} f(k) = \lim_{z \to 1}(z-1)F(z) \qquad (2.36)$$

In some texts, instead of the term $(z-1)$ multiplying F(z), we see the term $(1-z^{-1})$. But these are equivalent in the limit as $z \to 1$.

Note that in Equation 2.36, if f(∞) is to have a finite value, unstable modes must not be present, meaning that all poles of F(z) must lie inside or on the unit circle in the z-plane. The notion of *stability*, defined more formally toward the end of the chapter, is an important consideration in both analog and digital systems. Briefly, if f(∞) is to be finite, f(k) must not grow without bound as $k \to \infty$. The denominator of F(z), consisting say of terms such as $(z+a)(z-b)$, gives rise to time-sequence terms $(-a)^k$ and $(b)^k$ and for these to diminish as $k \to \infty$ requires that $|a| \leq 1$ and $|b| \leq 1$. But -a and b are poles of F(z). The poles of F(z), then, must lie inside or on the unit circle in the z-plane. (We assume that the reader is familiar with the notions of pole and zero from continuous-time systems.)

Example 2.17

Find the final value of a sequence that has the z-transform

$$F(z) = \frac{z^2}{z^2 - 1}$$

Solution: We can use the final value theorem to obtain

$$f(\infty) = \lim_{k \to \infty} f(k) = \lim_{z \to 1}(z - 1)F(z)$$

$$= \lim_{z \to 1}\left(\frac{z^2}{z^2 - 1}\right)(z - 1) = \lim_{z \to 1}\frac{(z^2)(z - 1)}{(z - 1)(z + 1)} = \frac{1}{2}$$

The initial and final value theorems are often useful if F(z) is known, but they are too complicated to obtain f (k) easily via the inverse z-transform. At least we can gather some information about the time sequences. The theorems are especially valuable if knowledge of f(k) at k = 0 and k = ∞ is all that we need.

Theorem 2.6 (Time Shift Theorem). If F(z) is the z-transform of a sequence f(k), then for the f(k-n) sequence, which is the f(k) sequence shifted n time units (n > 0) to the *right*, the corresponding z-transform is $z^{-n}F(z)$. This assumes that our sequences are causal and the z-transform is one-sided.
   If f(k) ↔ F(z), then for the f(k + m) sequence, which is the f(k) sequence shifted m time units (m > 0) to the *left*, the corresponding z-transform is $z^m F(z) - z^m f(o) - \cdots - zf(m - 1)$ and we can write

$$f(k - n) \leftrightarrow z^{-n}F(z) \qquad (2.37)$$

and

$$f(k+m) \leftrightarrow z^m F(z) - z^m f(o) - \cdots - zf(m-1) \tag{2.38}$$

There is one other case to consider. If f(k) has initial conditions associated with it, as often is the case in difference equation expressions, we can assume that f(-1), f(-2), and so on, are not necessarily zero. In this case

$$f(k-n) = z^{-n} F(z) + \sum_{x=1}^{n} z^{-n+x} f(-x) \tag{2.39}$$

The two most often encountered expressions which will be useful in solving difference equations are

$$f(k-1) \leftrightarrow z^{-1} F(z) + f(-1) \tag{2.40}$$

and

$$f(k-2) \leftrightarrow z^{-2} F(z) + f(-2) + z^{-1} f(-1) \tag{2.41}$$

Example 2.18

Given f(k) = 5ku(k), use the time-shift properties of the z-transform to find $Z\{f(k+1)\}$ and $Z\{f(k-1)\}$.

Solution: We use the Table 2.1 of z-transform pairs to find

$$F(z) = \frac{5z}{(z-1)^2}$$

then use the time-shift properties of z-transforms to obtain

$$Z\{f(k+1)\} = zF(z) - zf(0) = \left[\frac{5z^2}{(z-1)^2} \right]$$

$$Z\{f(k-1)\} = z^{-1} \left[\frac{5z}{(z-1)^2} \right] = \left[\frac{5}{(z-1)^2} \right]$$

assuming zero initial conditions.

Example 2.19

If $f(k) \leftrightarrow F(z)$ with $f(k) = (0.5)^k u(k)$ determine the z-transform of $f(k + 3) = (0.5)^{k+3} u(k + 3)$ and the z-transform of $f(k-3)$, assuming that $f(-1) = 1$, $f(-2) = 2$, and $f(-3) = 3$.

Solution: We know that

$$F(z) = \frac{z}{z - 0.5}$$

and

$$c_{zs} = h(k) * r(k)$$

Also,

$$Z\{f(k-3)\} = z^{-3}F(z) + z^{-2}f(-1) + z^{-1}f(-2) + f(-3)$$

$$= \frac{z^{-2}}{z - 0.5} + z^{-2} + 2z^{-1} + 3 = \frac{0.5z^{-2} + 3z + 0.5}{z - 0.5}$$

Example 2.20

Determine the z-transform of $(-0.5)^k u(k - 5)$, assuming zero initial conditions.

Solution: We know that

$$(-0.5)^k u(k) \leftrightarrow \frac{z}{z + 0.5}$$

and

$$(-0.5)^{k-5} u(k - 5) \leftrightarrow \frac{z^{-4}}{z + 0.5}$$

but this is not quite what is asked for. We need to manipulate our time sequence into the desired form: Multiply by $(-0.5)^5$ and we get

$$(-0.5)^k u(k-5) \leftrightarrow \frac{-1}{32} \frac{z^{-4}}{(z+0.5)}$$

Some of these shifting problems are not so easy to unravel. If f(k), for instance, is $k^2 u(k-2)$, this can be written as

$$f(k) = (k-2)^2 u(k-2) + 4(k-2)u(k-2) + 4u(k-2)$$

and each of these terms is easily z-transformable. But if f(k) is $\cos(\frac{\Pi}{3}k)u(k-2)$, some tedious trigonometric manipulations are required to reduce the expression to easily transformable terms, and to get the z-transform it might be easier just to use the defining summation.

To motivate the inverse z-transform, a few more examples that employ some of the z-transform properties and point to the need for the inverse z-transform will now be considered.

Example 2.21

Assume that we have a linear system with input r(k)= $(-0.5)^k u(k-5)$ and that the unit pulse response of the system is h(k)= $(-0.5)^k u(k)$. Determine the zero-state response of the system.

Solution: We know that

$$R(z) = \frac{-1}{32} \frac{z^{-4}}{(z+0.5)}$$

from Example 2.20. Also, H (z) $= \dfrac{z}{z+0.5}$ and the convolution $c_{zs} = h(k)*r(k)$ becomes $c_{zs}(z) = H(z)R(z)$ after taking the z-transform. Thus

$$c_{zs}(z) = \frac{1}{32} \frac{z^{-3}}{(z+0.5)^2}$$

To get back to the time domain we need the inverse z-transform, but $c_{zs}(k)$ can be obtained here just by using known transform pairs and properties of the z-transform. We know that

$$\frac{z}{(z+0.5)^2} \longleftrightarrow k(-0.5)^{k-1}u(k)$$

Multiply the left-hand side (LHS) by z^{-4} and since $c_{zs}(k)$ is assumed to be causal we get $(k-4)(-0.5)^{k-5}$ u (k - 4) on the RHS. Therefore,

$$c_{zs}(k) = \frac{-1}{32}(k-4)(-0.5)^{k-5}u(k-4)$$

Example 2.22

Use the z-transform to solve this difference equation:

c(k - 1) - 2 c(k) = r(k)

where $r(k) = 2\,(-1)^k\,u(k)$ and c(-1) = 1.5.

Solution: Taking the z-transform yields

$$\{z^{-1}C\,(z) + c\,(-1)\} - 2\,C\,(z) = R\,(z) = \frac{2z}{z+1}$$

and

$$C(z)\{z^{-1} - 2\} = -1.5 + \frac{2z}{z+1}$$

or

$$C(z) = \frac{(0.75 - 0.25z)z}{(z+1)(z-0.5)}$$

This function must be inverse z-transformed to get c(k). The problem is that known transform pairs and properties cannot simply be employed here. We need a more systematic approach to the inverse z-transform.

Example 2.23

Consider the system represented in the block diagram of Fig. 2.11 and determine c(k) if the input r(k) = δ(k) and there are zero initial conditions on the delay devices.

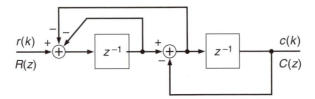

Figure 2.11   Block diagram for Example 2.23.

Solution: The delay devices can be viewed as gains of z^{-1} if all the signals are z-transformed. This is because q(k) at an input yields q(k - 1) at the output of a delay device and the z-transform yields Q(z) at the input and z^{-1} Q(z) at the output: The delay device multiplies the input to it by a factor of z^{-1}. Let the inputs of the delay devices from left to right be a $a_1(k) \leftrightarrow A_1(z)$ and $a_2(k) \leftrightarrow A_2(z)$; let r(k) = δ(k) $\leftrightarrow$ R(z) = 1 and c(k) $\leftrightarrow$ C(z). Then

$$C(z) = z^{-1}A_2(z)$$

and

$$A_2(z) = z^{-1}A_1(z) - C(z)$$

and

$$A_1(z) = R(z) - z^{-1}A_1(z) - A_2(z)$$

Let R(z) = 1 and solve for C(z)

$$zC(z) = A_2(z)$$

and

$$A_1(z)(1 + z^{-1}) = 1 - A_2(z) = 1 - zC(z)$$

so

$$A_1(z) = \frac{1 - zC(z)}{1 + z^{-1}}$$

and

$$zC(z) = z^{-1}\{\frac{1 - zC(z)}{1 + z^{-1}}\} - C(z)$$

Then C (z) = $1 / (z^2 + 3z + 1)$ and to get c (k) we again need more than just z-transform properties. We need an inverse z-transform procedure, which we will consider after a summary of z-transform properties.

Table 2.2 summarizes the z-transform properties.

Table 2.2 Properties of the z-transform

f(k)	F(z)
$\alpha f(k)$	$\alpha F(z)$
$f_1(k) + f_2(k)$	$F_1(z) + F_2(z)$
$Z\{f(k)*g(k)\}$	$F(z)G(z)$
f (0)	$\lim_{z \to \infty} F(z)$
$f(\infty)$	$\lim_{z \to 1}(z - 1)F(z)$
$f(k + m)$	$z^m F(z) - z^m f(0) - \cdots - zf(m - 1)$
$f(k - n)$	$z^{-n}F(z)$ If f (k) is purely causal
$f(k - n)$	$z^{-n}F(z) + \sum_{x=1}^{n} z^{-n+x} f(-x)$ if f (k) has initial conditions

Inverse z-transform

The inverse z-transform is the process of obtaining the discrete-time signal f(k) whose z-transform F(z) is given, where we assume that f(k) and F(z) are z-transform pairs. Several methods can be used to find the inverse z-transform, including complex integration, power-series expansion, and partial-fraction expansion. In this book we discuss the partial-fraction expansion method, the method of choice of most engineers for evaluating an inverse z-transform and also the method familiar to most engineers who have a background in Laplace transforms.

To find the original discrete signal f(k) from its z-transform F(z) using the partial-fraction expansion method, expand F(z) into a summation of terms each of which has an inverse z-transform that is readily recognizable from z-transform tables (e.g., Table 2.1). In Appendix D we explain the details of how to expand functions in partial fractions. Appendix C contains additional z-transform tables. CAE packages are available to evaluate the partial-fraction expansion of a function. These packages also calculate the inverse z-transform of a function (see Appendix A).

One thing to keep in mind about partial-fraction expansions for the z-transform (unlike for the Laplace transform) is that generally simpler results are obtained if instead of expanding F(z) we expand F(z)/z. The next example illustrates this idea.

Example 2.24

Find the inverse z-transform of the function

$$F(z) = \frac{z}{(z-1)(z-10)}$$

Solution: Using the partial-fraction expansion method in Appendix D, we obtain

$$\frac{F(z)}{z} = \frac{1}{(z-1)(z-10)}$$

$$= \frac{(1-10)^{-1}}{z-1} + \frac{(10-1)^{-1}}{z-1}$$

$$= \frac{-1}{9(z-1)} + \frac{1}{9(z-10)}$$

Then cross-multiply by z and we obtain

$$F(z) = -\frac{1}{9}\frac{z}{z-1} + \frac{1}{9}\frac{z}{z-10}$$

which are two terms readily transformable using the z-transform pair

$$a^k u(k) \leftrightarrow \frac{z}{z-a}$$

Therefore,

$$f(k) = \{-1/9 + 1/9(10)^k\}u(k)$$

Now, if we expand just F(z), we get

$$F(z) = \frac{A}{z-1} + \frac{B}{z-10} = \frac{-\frac{1}{9}}{z-1} + \frac{\frac{10}{9}}{z-10}$$

and to invert these terms we need to invoke the time-shift property. Multiply by z, then z^{-1}, and we get

$$F(z) = -\frac{1}{9}(\frac{z}{z-1})z^{-1} + \frac{10}{9}(\frac{z}{z-10})z^{-1}$$

implying

$$\frac{-1}{9}u(k) + \frac{10}{9}10^k u(k)$$

which needs to be shifted to get

$$f(k) = \frac{-1}{9}u(k-1) + \frac{10}{9}10^{k-1}u(k-1)$$

$$= \frac{-1}{9}u(k-1) + \frac{1}{9}10^k u(k-1)$$

This matches our previous result since f(0) = 0. The point is that the first approach, expanding $F(z)/z$, generally avoids the shifting property.

The case of repeated roots, of course, complicates the process of partial fraction expansion. But again, the approach is similar to the Laplace transform approach.

Example 2.25

Find f(k) if

$$F(z) = \frac{z^4}{(z-1)(z-0.5)^3}$$

Solution:

$$\frac{F(z)}{z} = \frac{A}{z-1} + \frac{B}{z-0.5} + \frac{C}{(z-0.5)^2} + \frac{D}{(z-0.5)^3}$$

The D and A terms are obtained directly: $A = 8$ and $D = -1/4$. There are several ways to get B and C, including the use of a differentiation formula, cross-multiplying and equating coefficients, or letting z take on values (not 1 or 0.5). Letting z = 1.5 and z = 2, we get B = -7 and C = -2. Then

$$F(z) = \frac{8z}{z-1} - \frac{7z}{z-0.5} - 2\frac{z}{(z-0.5)^2} - \frac{1}{4}\frac{z}{(z-0.5)^3}$$

Using the z-transform pairs

$$\frac{z}{z-a} \leftrightarrow a^n \; ; \; \frac{z}{(z-a)^2} \leftrightarrow na^{n-1} \; ; \; \frac{z}{(z-a)^3} \leftrightarrow \frac{n(n-1)}{2!}a^{n-2}$$

we get

$$f(k) = 8 - 7(0.5)^k - 2k(0.5)^{k-1} - \frac{0.25}{2}k(k-1)(0.5)^{k-2}$$

$$= \{8 - (0.5)^k(7 + 3.5k + 0.5k^2)\}u(k)$$

MATLAB has a statement that is useful for partial-fraction expansion in the inverse z-transform: [R, P, K]= RESIDUE (B, A). All we need to do is enter F (z) = B (z)/A (z) via two statements, A=[...] and B=[...]. A is the vector of denominator coefficients, B the vector of numerator coefficients. The coefficients are entered into A and B in ascending powers of z^{-1} terms starting with constants. An example follows.

Example 2.26

Given F (z) = $\dfrac{2z}{z^2 + 3z + 2}$ find F (k).

Solution: Let

$$F(z) = \frac{2z^{-1}}{1 + 3z^{-1} + 2z^{-2}}$$

Then A=[1 3 2] and B=[0 2] and the string of statements

      A =[1 3 2]
      B =[0 2]
      [R, P, K] = RESIDUEZ (B, A)

returns

$$R = \begin{bmatrix} -2 \\ 2 \end{bmatrix}, P = \begin{bmatrix} -2 \\ 1 \end{bmatrix}, K = (0)$$

K is a vector of terms to be added to F (z) if the order of the numerator is greater than the order of the denominator. Our F(z) is expanded into a sum of terms of the form $r/(1 - pz^{-1})$ where r is the residue and p the pole. So we get

$$F(z) = \frac{-2}{1 + 2z^{-1}} + \frac{2}{1 - z^{-1}} = \frac{-2z}{z + 2} + \frac{2z}{z - 1}$$

and these are in easily invertible form. Therefore,

$$f(k) = (-2(-2)^k + 2(-1)^k)u(k)$$

2.5 DISCRETE SYSTEM TRANSFER FUNCTION

A system transfer function is used to analyze both discrete and continuous systems. It contains important information on a system's stability and response characteristics. The locations of poles and zeros of a system transfer function govern these characteristics. Fig. 2.12 shows the block diagram of a general discrete system transfer function, H (z), where

$$H(z) = \frac{C(z)}{R(z)} \tag{2.42}$$

or

$$C(z) = R(z)H(z) \tag{2.43}$$

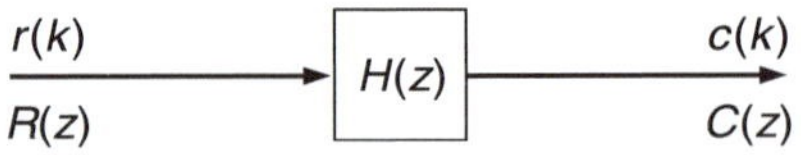

Figure 2.12   Discrete system transfer function.

This equation also follows from taking the z-transform of the equation

$$c_{zs}(k) = r(k)*h(k) \tag{2.44}$$

Once the system transfer function is known, Equation 2.43 is used to find the z-transform of the system response.

An advantage of using the system transfer function representation of a system is that we can find the response of a linear system to various inputs, as long as we assume a *quiescent* system: All initial conditions are zero. The response to an impulse, the *impulse response*, h(k), is just the inverse z-transform of H(z) because R(z) = 1 when r(k) = $\delta(k)$. How a discrete system responds to a unit step, the *step response* of the system, is often also an important determination. With R(z) = $z/(z-1) \leftrightarrow$ u(k) = r(k), it follows that

$C(z) = \dfrac{z}{(z-1)} H(z)$ where $C(z) \leftrightarrow c(k)$ and c(k) is the step response.

There is a MATRIX$_X$ statement that provides a numerical solution for c(k) when r(k) = u(k):

<N, Y> = DSTEP (DNUM, DDEN, NPTS)

DNUM is the vector of numerator coefficients of H(z) (in order of descending powers of z), DDEN is the vector of denominator coefficients, and NPTS is the number of response time points.

Example 2.27

Let a system be described by

$$H(z) = \frac{z^4 + 2z^3 + 3z^2 + 2z + 1}{z^4 + z^3 + z^2 + z + 1}$$

and plot the unit step response.

Solution: We enter the coefficients and the following string of statements:

> DNUM = [1 2 3 2 1]
> DDEN = [1 1 1 1 1]
> NPTS = 20
> <N, Y> = DSTEP (DNUM, DDEN, NPTS)

and MATRIX$_X$ produces the points, which can be plotted as shown in Fig. 2.13.

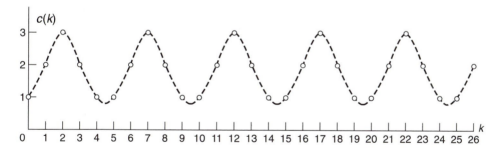

Figure 2.13   Plot of a step response using MATRIX$_X$.

Note that the response c(k) is oscillatory. To check the roots of the DDEN polynomial (system poles) we can use the MATRIX$_X$ statements ROOTS(x), where we let x = DDEN. It returns the values

> Z = -0.8090 + 0.5878j
> Z = -0.8090 - 0.5878j

$$Z = 0.3090 + 0.9511j$$
$$Z = 0.3090 - 0.9511j$$

all four of which have magnitudes of 1.0 and indicate poles of H(z) on the unit circle in the z-plane, which implies behavior between a state of growing without bound and decaying exponential-like behavior, or oscillatory behavior, in the discrete-time domain.

Often, we will need to deal with initial conditions in our systems. Are transfer functions of any use with these situations? Yes, at least indirectly. Once H(z) is known, the system difference equation can be constructed by cross multiplying and letting $R(z) \leftrightarrow r(k)$, $C(z) \leftrightarrow c(k)$, and $z^{-1}R(z) \leftrightarrow r(k-1)$, $z^{-1}C(z) \leftrightarrow c(k-1)$, and so on. Then the zero-input response or the total response, or even the zero-state response can be obtained from either the z-transform or from classical time-domain solution methods.

Example 2.28

Determine the zero-state response and the zero-input response for a system with $H(z) = z/(z^2 + 2z + 1)$.

Solution: Assume that $r(k) = (-0.5)^k u(k)$ and $c(-1) = c(-2) = 2.0$.

$$H(z) = \frac{C(z)}{R(z)} = \frac{z^{-1}}{1 + 2z^{-1} + z^{-2}}$$

Then

$$C(z) + 2C(z)z^{-1} + C(z)z^{-2} = z^{-1}R(z)$$

or

$$c(k) + 2c(k-1) + c(k-2) = r(k-1)$$

To get the zero input response, get the characteristic equation $r^2 + 2r + 1 = 0$. Then

$$c_{zi}(k) = \alpha_1(-1)^k + \alpha_2 k(-1)^k$$

and using $c(-1) = c(-2) = 2$, we get $\alpha_1 = -6$ and $\alpha_2 = -4$. Therefore,

$$c_{zi}(k) = (-4 - 6k)(-1)^k u(k)$$

To get the zero state response write

$$C_{zs}(z) = H(z)R(z) = \frac{z}{(z+1)^2} \frac{z}{z+0.5}$$

and

$$\frac{C_{zs}(z)}{z} = \frac{z}{(z+1)^2(z+0.5)} = \frac{\alpha}{z+0.5} + \frac{\beta}{z+1} + \frac{\gamma}{(z+1)^2}$$

Thus $\alpha = -2$, $\beta = 2$, $\gamma = 2$ and

$$c_{zs}(k) = \{-2(-0.5)^k + 2(-1)^k - 2k(-1)^k\}u(k)$$

There are three primary types of connections of systems and thus of transfer functions representing systems: (a) cascade, (b) parallel, and (c) canonical. These are illustrated in Fig. 2.14.

(a)

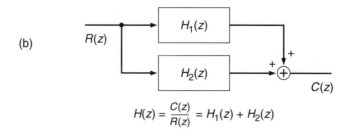

$$H(z) = \frac{C(z)}{R(z)} = H_1(z)H_2(z)$$

(b)

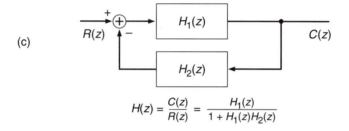

$$H(z) = \frac{C(z)}{R(z)} = H_1(z) + H_2(z)$$

(c)

$$H(z) = \frac{C(z)}{R(z)} = \frac{H_1(z)}{1 + H_1(z)H_2(z)}$$

Figure 2.14 (a) Cascaded transfer functions; (b) parallel transfer functions; (c) canonical transfer functions.

A typical problem requires us to determine an over all transfer function, H (z) = $C(z)/R(z)$, when the system between r(k) and c(k) consists of several subsystems of the aforementioned three types. The next example illustrates this scenario.

Example 2.29

Determine $C(z)/R(z)$ for the discrete control system shown in Fig. 2.15a using block diagram reduction.

Solution: See Fig. 2.15b.

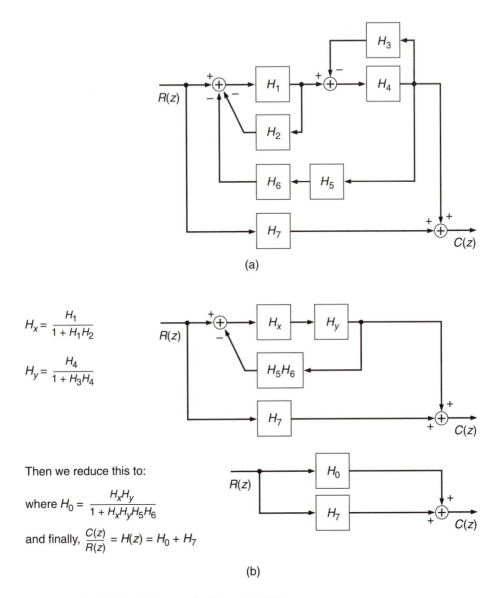

$$H_x = \frac{H_1}{1 + H_1 H_2}$$

$$H_y = \frac{H_4}{1 + H_3 H_4}$$

Then we reduce this to:

where $H_0 = \dfrac{H_x H_y}{1 + H_x H_y H_5 H_6}$

and finally, $\dfrac{C(z)}{R(z)} = H(z) = H_0 + H_7$

(b)

Figure 2.15 Block diagram for Example 2.29.

The Mason gain rule provides a systematic way to perform block-diagram reduction by isolating loop gains and transmittances and distinguishing between touching and nontouching loops. We do not pursue this method here but refer the reader to classical controls texts (e.g., [8]).

Time-domain results are often needed when transfer functions are given. In the *cascade* connection we can write in the time domain

$$C(z)=R(z) H(z) \tag{2.45}$$

where

$$H(z) = H_1(z)H_2(z) \tag{2.46}$$

$$h(k) = h_1(k)*h_2(k) = \sum_{n=0}^{k} h_1(n)h_2(k-n) \tag{2.47}$$

and

$$C(z) = R(z)H(z) \tag{2.48}$$

so

$$c(k) = r(k)*h(k) = \sum_{n=0}^{k} r(n)h(k-n) \tag{2.49}$$

Given $H_1(z)$ and $H_2(z)$, we would solve Equation 2.47 first, then solve Equation 2.49. In the *parallel* connection we can write in the time domain

$$c(k) = r(k)*(h_1(k)+h_2(k)) = \sum_{n=0}^{k} r(n)\{h_1(k-n)+h_2(k-n)\} \tag{2.50}$$

and in the *canonical* connection we let

$$E(z) = R(z) - H_2(z)C(z) \tag{2.51}$$

so

$$e(k) = r(k) - \sum_{n=0}^{k} h_2(n)c(k-n) \tag{2.52}$$

and

$$E(z)H_1(z) = C(z) \qquad\qquad (2.53)$$

so

$$c(k) = \sum_{n=0}^{k} e(n)h_1(k - n) \qquad\qquad (2.54)$$

In this case Equations 2.52 and 2.54 need to be solved simultaneously.

The general form of a discrete transfer function is

$$H(z) = \frac{b_0 + b_1 z^{-1} + \cdots b_m z^{-m}}{1 - a_1 z^{-1} - a_2 z^{-2} - \cdots - a_n z^{-n}} \qquad\qquad (2.55)$$

where $n \geq m$. By multiplying both the numerator and the denominator by z^n we arrive at

$$H(z) = \frac{b_0 z^n + b_1 z^{n-1} + \cdots b_m z^{n-m}}{z^n - a_1 z^{n-1} - a_2 z^{n-2} - \cdots - a_n z^0} \qquad\qquad (2.56)$$

These types of transfer functions are central to all of what follows.

2.6 FREQUENCY RESPONSE

The transfer functions represented in Equations 2.55 and 2.56 reveal much information about the systems from which they arise. As indicated earlier, the pole locations indicate system stability. Another key idea associated with transfer functions is the *frequency response* of a system. We assume that the reader is familiar with the basic ideas associated with frequency response of an analog system represented by H(s). The frequency response of a discrete system, represented by the discrete transfer function H(z), is also crucial to the design and analysis of discrete-time systems.

To determine the frequency response of a transfer function H(z) we need to let z vary over some range. But which range? From H(s) we recall that $s = j\omega$ was required to obtain the frequency response and we varied ω from minus infinity to plus infinity, which corresponded to running up the middle of the s-plane from bottom to top. But the $j\omega$-axis also separated stable from unstable regions of the s-plane. We indicated earlier (following the discussion of the final value theorem) that the unit circle in the z-plane separated stable and unstable regions for the discrete-time systems. Thus the frequency response of H(z) can be obtained by plotting the magnitude and phase of H(z) evaluated at

$$z = e^{j\theta} \tag{2.57}$$

where $-\pi \le \theta \le \pi$.

Example 2.30

Plot the magnitude and phase of the frequency response of the system shown in Fig. 2.16.

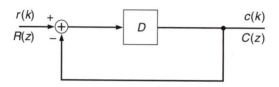

Figure 2.16   System of Example 2.30.

Solution: The output of the summer is

$$c(k + 1) = r(k) - c(k)$$

and the z-transform of this expression is

$$zC(z) = R(z) - C(z)$$

from which we get

$$H(z) = \frac{C(z)}{R(z)} = \frac{1}{z+1}$$

Then letting $z = e^{j\theta}$, we can write

$$H(e^{j\theta}) = \frac{1}{\cos\theta + j\sin\theta + 1}$$

from which we get the magnitude and phase

$$|H| = \frac{1}{\sqrt{2}} \frac{1}{\sqrt{1 + \cos\theta}}$$

and

$$\arg H = -\tan^{-1}\left(\frac{\sin\theta}{1 + \cos\theta}\right)$$

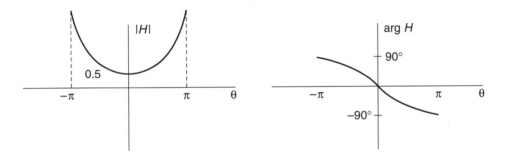

Figure 2.17   Magnitude and phase plots for Example 2.30.

The plots (Fig. 2.17) are periodic with period 2π, and for H(z)'s describing real systems, magnitude plots are *even* functions of θ and phase plots are *odd*. So only the range of θ from zero to π is needed.

There is a MATLAB statement that provides frequency response for transfer functions of the form

$$H(z) = \frac{B(z)}{A(z)} \tag{2.58}$$

It takes the form

$$[H, W] = FREQZ (B, A, N) \tag{2.59}$$

H and W are magnitude and phase. A is the vector of denominator coefficients, B the vector of numerator coefficients. The coefficients are entered into A and B in ascending powers of

z^{-1} terms starting with constants. N is the number of points we want to plot in the range from zero to π. Using this statement with Example 2.30 to get a plot, we need only type

> A=[1 1]
> B=[0 1]
> FREQZ (B, A)

(The left-hand side of the expansion [H,W] can be suppressed if we only want plots.) MATLAB will return the plots shown in Fig. 2.18.

We do not need to enter a value for N. The program defaults to 512 points.   Also note that the magnitude is in decibels and the frequency is normalized so that $\theta = \pi$ corresponds to 1.0.

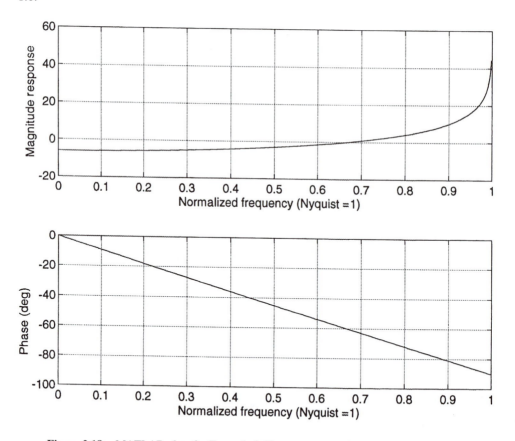

Figure 2.18   MATLAB plots for Example 2.30.

Example 2.31

Plot the frequency response for

$$H(z) = \frac{z^2 + z + 1}{z^4 - 0.8125z^2 + 0.1406}$$

Solution: Write H(z) as follows:

$$\frac{z^{-2} + z^{-3} + z^{-4}}{1 - 0.8125z^{-2} + 0.1406z^{-4}}$$

Enter B = [0 0 1 1 1] and A = [1 0 -0.9125 0 0.1406]. MATLAB produces the plots indicated in Fig. 2.19.

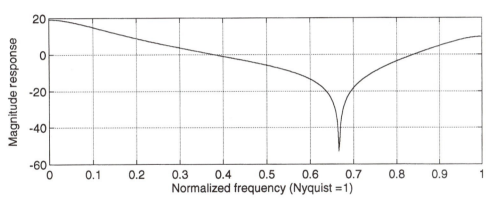

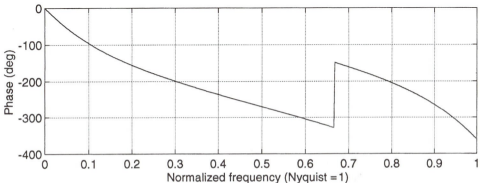

Figure 2.19 MATLAB plots for Example 2.31.

Frequency response will be an essential tool in discrete-time control systems design and analysis procedures. Numerical tools, such as MATLAB with its FREQZ statement are recommended to perform frequency response of discrete-time systems. For continuous-time systems we often use the Bode representations for frequency response of $H(s)|_{s=j\omega} = H(j\omega)$. The ease of Bode construction, knowing the poles and zeros of H(s), makes this method attractive. But because z is not linearly related to ω (as is s in the relation $s = j\omega$), to plot $H(z) = |H(z)|$ with $z = e^{j\theta}$ and $\theta = \omega T$ using the Bode technique without machine computation is prohibitive. One thing we can do in this case is to map the z-plane into a w-plane via the relation

$$w = \frac{1}{T}\ln z \qquad (2.60)$$

approximated by

$$w = \frac{2}{T}(\frac{z-1}{z+1}) \qquad (2.61)$$

and

$$z = e^{j\omega T}$$

yields

$$w = j\frac{2}{T}\tan\frac{\omega T}{2} \qquad (2.62)$$

So when z takes on values around the unit circle, w takes on values along the imaginary axis in the w-plane. This behavior in the w-plane is very similar to behavior in the s-plane and the frequency response of H(w) can be sketched using the Bode straight-line approximations.

In addition to frequency response of *systems*, the frequency content or frequency spectrum of *signals* is occasionally important. For example, in systems with a nonlinear element, the signal leaving the nonlinearity may be input to a plant that acts like a filter to remove the higher harmonics in the signal. Here the spectral content of the signal is crucial, especially if the technique called *describing function* (DF) *analysis* is employed. DF analysis is discussed in most books on nonlinear control systems (cf. [10]).

For nonperiodic signals, the signal frequency content is revealed most directly by the *Fourier Transform*. Fourier series expansions are useful for periodic signals, but real physical signals are never strictly periodic. We must begin a measurement of, for example, voltage or position, at some point in time and stop at a later finite time; that is, our signals are always time limited. Hence although some signals can be approximated as periodic signals and do benefit from Fourier series analysis, the Fourier transform is employed more frequently by most practicing engineers. The Fourier transform $F(j\omega)$ of a signal f(t) is defined as

$$F(j\omega) = \int_{-\infty}^{\infty} f(t)e^{-j\omega t}\, dt \qquad\qquad (2.63)$$

and the corresponding inverse Fourier transform is

$$f(t) = \frac{1}{2\pi}\int_{-\infty}^{\infty} F(j\omega)e^{j\omega t}\, d\omega. \qquad\qquad (2.64)$$

As with the z-transform and the Laplace transform, we can indicate these with the double-arrow notation,

$$f(t) \Leftrightarrow F(j\omega) \qquad\qquad (2.65)$$

But since our primary concern is with discrete signals, not continuous signals, we will not deal directly with $F(j\omega)$ and $f(t)$, but instead, concern ourselves with a *discrete Fourier transform* (DFT) and *inverse discrete Fourier transform* (IDFT).

The DFT can be developed from the z-transform. The z-transform of f(n) is defined as

$$F(z) = \sum_{n=0}^{\infty} f(n)z^{-n} \qquad\qquad (2.66)$$

In the discussion of frequency response of F(z) or H(z), we let z be $e^{j\theta}$ and varied θ from $-\pi$ to π or, equivalently, from 0 to 2π. If this is the case in Equation 2.66, we get

$$F(e^{j\theta}) = F(\theta) = \sum_{n=0}^{\infty} f(n)e^{-jn\theta} \qquad\qquad (2.67)$$

$F(\theta)$ is sometimes called the *discrete-time Fourier transform* and is almost, but not quite, the DFT.

Since the discrete signals we deal with are real, they are indicated by a finite, not an infinite, number of points. That requires us to limit the summation in Equation 2.67 to a finite number of points. Let us assume that we have N points, where for convenience we will let N be an even number. Also discretize θ by letting $\theta = 2\pi k/N$ for k = (0, 1, 2,..., N-1). Then we can write

$$F(k) = \sum_{n=0}^{N-1} f(n)e^{-j(\frac{2\pi}{N})kn} \tag{2.68}$$

and by writing this equation out for k = 0, 1, 2,..., N-1, collecting terms, and inverting a matrix, we can write

$$f(n) = \frac{1}{N}\sum_{k=0}^{N-1} F(k)e^{j(\frac{2\pi}{N})kn} \tag{2.69}$$

These are, respectively, the DFT and the IDFT. More concise expression is available by letting

$$e^{-j(\frac{2\pi}{N})} = \eta. \tag{2.70}$$

Then

$$F(k) = \sum_{n=0}^{N-1} f(n)\eta^{kn} \tag{2.71}$$

and

$$f(n) = \frac{1}{N}\sum_{k=0}^{N-1} F(k)\eta^{-kn} \tag{2.72}$$

and again we can write

$$f(n) \Leftrightarrow F(k) \tag{2.73}$$

Since the mechanisms involved in the F(k) and f(n) expressions are so similar, we can use the same computational algorithm to compute F(k) and f(n). All we need to do is conjugate η^{kn} to get η^{-kn} and multiply by the 1/N factor to get f(n) from the algorithm that computes F(k).

Now, even though real f(n) functions are finite in duration and therefore are nonperiodic, the DFT and IDFT computations are periodic mechanisms. Because of this periodicity, there is inevitable overlap, called *aliasing*, at the edges of the F(k) spectrum. If N is large enough, the ill effects of aliasing are generally mitigated. The key to a large N is to sample the data record as fast as possible. There are also several clever ways to minimize aliasing using filters, but these are beyond our present scope.

Another DFT problem is *leakage*. This problem arises from the necessary abruptness of starting and stopping the sampling of a data record. The frequency spectra of such data records generally consist of main lobes which contain most of the spectral information but also sidelobes in which information is lost. This lost information is referred to as leakage. Leakage can be reduced by sidelobe minimization procedures. Several such procedures have been developed in the signal processing literature. Most employ *windowing*, a technique used to smooth the abruptness of the discontinuities in the data records. Many different kinds of windows have been developed in the last two decades. Although the details of windowing are beyond our present scope, a glance at the literature should convince the reader that the selection of the best window for a given data record is often more an art than a science.

To compute the DFT and/or the IDFT from Equations 2.71 and 2.72, we need to perform a great number of additions and multiplications if N is large, for example, if N = 256. Computer solutions are naturally called for. In addition, advantage can be taken of the symmetries and periodicities involved in the DFT/IDFT equations. This is precisely what the fast fourier transform (FFT) does. The FFT is an algorithm that computes the DFT with far fewer additions and multiplications than are called for in Equations 2.71 and 2.72. A great savings in computer time can be had with the FFT, especially for large N. Many different algorithms have been developed, among the most popular being the decimation-in-time and the decimation-in-frequency algorithms. The CAE packages MATLAB and MATRIX$_X$ — our preferred computational tools — do employ FFT algorithms to compute the DFT and IDFT.

Example 2.32

Using MATLAB, compute the DFT for the data record x = [1 1 1 1 1 1 1 1].

Solution: All we need to do is enter x as given and write DFT(x) and MATLAB returns [8 0 0 0 0 0 0 0]. The reader is invited to verify that a longhand computation using Equation 2.71 produces the same result. MATRIX$_X$ will yield the same result with FFT (x).

2.7 RELATIONSHIP BETWEEN THE S AND Z DOMAINS

Understanding the relationship between continuous and discrete systems requires consideration of how the s-plane and z-plane are related. This is because many designs are developed using classical methods within the continuous domain and are then transformed into the discrete domain. The s- and z-planes are complex two-dimensional spaces related by the simple equation

$$z = e^{sT} \qquad (2.74)$$

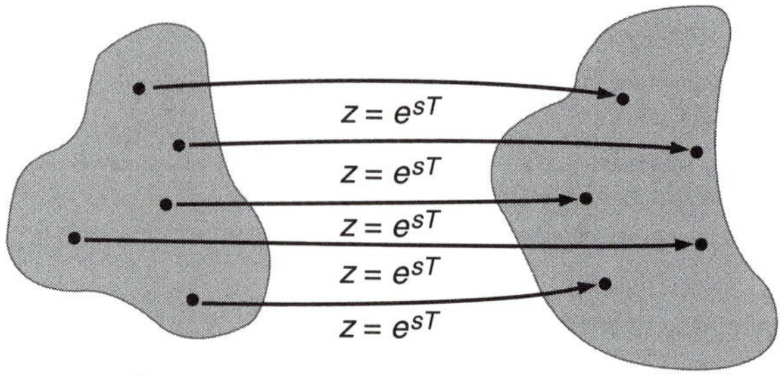

Figure 2.20   Mapping from the s- to the z-plane.

As indicated in Fig. 2.20, each point in the complex s-plane is mapped to a complex point in the z-plane. A point $s = a + jb$ in the s-plane maps to $z = e^{sT}$, where z is defined by a vector of length e^{at} and an angle bt. The mapping e^{sT} comes from sampling theory. If f(t) is sampled every T seconds to produce f(kT), $k = 0, 1, 2, \ldots$ and if the sampler closes for a very brief time $\varepsilon \ll T$, then in the limit as ε goes to zero we can assume that we have an ideal sampler whose output is $f^*(t)$. We can represent $f^*(t)$ by a train of impulses each occurring at the sampling points 0, T, 2T, 3T, ... and each impulse weighted with the value of f(t) at those points.

$$f^*(t) = \sum_{k=0}^{\infty} f(kT)\delta(t - kT) \qquad (2.75)$$

Then the Laplace transform of the sampled signal is

$$F^*(s) = \sum_{k=0}^{\infty} f(kT)e^{-skT} \qquad (2.76)$$

But this Laplace transform of the sampled version of f(t) is an exponential function of s and these are not convenient forms. To convert to a rational polynomial form, we let

$$s = \frac{1}{T} \ln z \tag{2.77}$$

to get

$$F^{*}(s)\Big|_{s=\frac{1}{T}\ln z} = \sum_{k=0}^{\infty} f(kT)z^{-k} \tag{2.78}$$

and this is exactly the form of the z-transform. Thus the $z = e^{sT}$ mapping relates s- and z-planes; in particular, the poles and zeros of the Laplace transform of f(t) relate to the poles and zeroes of the z-transform of f(kT) in accord with this mapping. Given the mapping rule $z = e^{sT}$, we can investigate how different regions of the s-plane map to the z-plane.

Region One: Left-Hand Plane

Each point in the left-hand plane (LHP) of the s-plane is mapped to a unit circle in the z-plane. This is shown in Fig. 2.21. From Fig. 2.21 we can establish an important rule, suggested earlier in our discussions, about the stability of a discrete-time system.

Theorem 2.7 (Stability Theorem). The stability boundary for discrete-time systems is the unit circle in the z-plane. This follows from $z = e^{sT}$ and if s = a + jb, $z = e^{aT}e^{jbT}$ where $e^{aT} = |z|$. But if $a \leq 0$, then $|z| \leq 1$. So points (e.g., poles) in the left half of the s-plane map into points interior to the unit circle in the z-plane. Poles of H(z) inside the unit circle correspond to stable discrete systems. Poles on $|z| = 1$ correspond to marginally stable systems. Repeated poles on $|z| = 1$ or poles outside the unit circle correspond to unstable discrete systems.

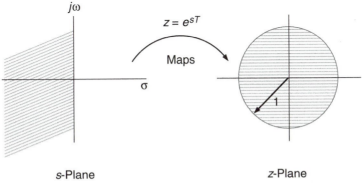

Figure 2.21 Mapping of the LHP of the s-plane to a unit circle of the z-plane.

Region Two: Right-Hand Plane

The right-hand plane (RHP) of the s-plane maps into the outside of the unit circle of the z-plane (Fig. 2.22). This follows from $z = e^{sT}$ and if $s = a+jb$, $z = e^{aT}e^{jbT}$ where $e^{aT} = |z|$. But if $a \geq 0$, then $|z| \geq 1$, which refers to points outside the unit circle.

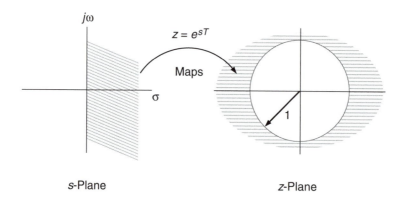

Figure 2.22   Mapping of the RHP of the s-plane to the region outside the unit circle of the z-plane.

Region Three: Imaginary Axis Boundary in the s-Plane

The imaginary axis boundary in the s-plane maps to the unit-circle boundary in the z-plane (Fig. 2.23). This follows from $z = e^{sT} = e^{j\omega T}$ which is the equation for a unit circle in the z-plane.

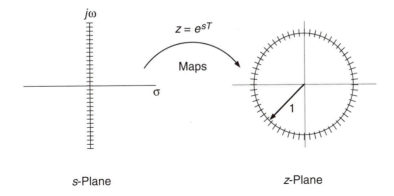

s-Plane   *z*-Plane

Figure 2.23   Mapping of the imaginary axis boundary in the s-plane to the unit circle boundary in the z-plane.

There are several other important mappings from the s-plane to the z-plane. For instance, if points in the s-plane have constant damping ratio corresponding to poles of the form $s^2 + 2\zeta\omega_n s + \omega_n^2 = 0$ or

$$s = -\zeta\omega_n + j\omega_n\sqrt{1-\zeta^2} \tag{2.79}$$

with ζ fixed and ω_n adjustable, then

$$z = e^{sT} = e^{-\zeta\omega_n T + j\omega_n T\sqrt{1-\zeta^2}} = re^{j\theta} \tag{2.80}$$

As ω_n varies from zero to infinity, z spirals into the origin, r diminishing from 1 to zero as θ starts at $0°$ and rotates clockwise. This behavior is indicated in Fig. 2.24.

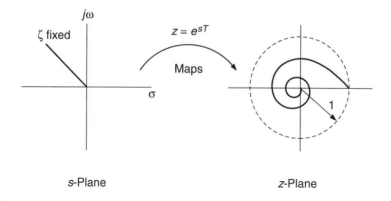

Figure 2.24   Mapping of a line along which are poles with constant damping ratio in the s-plane into a spiral in the z-plane.

There are many methods available for transforming an analog system transfer function, H(s), into a transfer function, H (z), that will describe a discrete system whose behavior approximates the analog system. We can begin our discussion with a consideration of integration. For an integrator, represented by $\frac{1}{s}$ with input $r(t) \leftrightarrow R(s)$ and output $c(t) \leftrightarrow C(s)$, we can write

$$r(t) = \frac{d}{dt} c(t) \tag{2.81}$$

$$c(t) = \int_{-\infty}^{t} r(\tau) d\tau$$

Since $r(t) \approx \dfrac{\Delta c}{\Delta T}$ and $\Delta T = T$ we can write

$$Tr(k) = c(k) - c(k-1) \tag{2.82}$$

$$c(k) = c(k-1) + Tr(k)$$

which is called *backward rectangular integration*. Taking the r(k-1) instead of r(k) value, we get

$$c(k) = c(k - 1) + Tr(k - 1) \qquad (2.83)$$

which is called *forward rectangular integration*. If instead of either r(k) or r(k - 1) we take their average, we get

$$C(z)/R(z) \qquad (2.84)$$

which is called *trapezoidal integration*.

The z-transform of these equations yields

$$\frac{C(z)}{R(z)} = \frac{Tz^{-1}}{1 - z^{-1}} = \frac{T}{z - 1} \qquad (2.85)$$

$$\frac{C(z)}{R(z)} = \frac{Tz}{z - 1} \qquad (2.86)$$

and

$$\frac{C(z)}{R(z)} = \frac{T}{2}(\frac{z + 1}{z - 1}). \qquad (2.87)$$

Since $C(s)/R(s) = 1/s$ these three expressions yield three approximations to s:

$$s = \frac{z - 1}{T} \quad \text{or } z = 1 + sT \qquad (2.88)$$

$$s = \frac{z - 1}{Tz} \quad \text{or } z = \frac{1}{1 - Ts} \qquad (2.89)$$

$$s = \frac{2}{T}(\frac{z - 1}{z + 1}) \quad \text{or } z = \frac{1 + \frac{T}{2}s}{1 - \frac{T}{2}s}. \qquad (2.90)$$

Since $z = e^{sT}$ and $s = \dfrac{1}{T} \ln z$ are unwieldy forms, the representations above serve as approximations that considerably simplify our transfer function manipulations.

The forward integration method is simple but maps the left half of the s-plane into all of the z-plane to the left of 1.0. That means that a stable H(s) system can have a corresponding H(z) that has poles outside the unit circle and is therefore unstable. The backward integration method is also simple but maps the left half of the s-plane into a circle contained in the right half of the unit circle of the z-plane. Such a mapping yields a very stable discrete system, but unfortunately much distortion also occurs. The trapezoidal integration affords the best approximation and, in fact, maps the entire left half of the s-plane into the inside of the unit circle in the z-plane. The use of trapezoidal integration to approximate s is sometimes called the *Tustin method* or the *Bilinear method*.

There is a MATRIX$_X$ statement that can do these basic conversions automatically:

<NUMD, DEND>=DISCRETIZ (NUM, DEN, DT,'TYPE').   (2.91)

DT is the value of T, and NUM and DEN are the numerator and denominator of a given H(s) expressed in descending powers of s. In the 'TYPE' position we can enter several types of s- to z-plane transformation, including

      'FORWARD'
      'BACKWARD'
      'TUSTIN'

The following example illustrates some of these.

Example 2.33

Let

$$H(s) = \frac{s+3}{s^2 + 3s + 2}$$

and using the Forward integration method with T = 1 and T = 0.1 find the corresponding H(z) equivalents. Compare results for the T = 1 case with the backward and Tustin methods.

Solution

      NUM=[1 3]
      DEN= [1 3 2]
      T=1

returns $\text{DEND} = \begin{bmatrix} 1 & 1 & 0 \end{bmatrix}$ and $\text{NUMD} = \begin{bmatrix} 1 & 2 \end{bmatrix}$
or

$$H(z) = \frac{z+2}{z^2+z}$$

and with $T = 0.1$ we get

$$H(z) = \frac{0.1(z-0.7)}{z^2 - 1.7z + 0.72}$$

In the backward case we get with $T = 1$

$$H(z) = \frac{0.67z^2 - 0.167z}{z^2 - 0.833z + 0.167}$$

and in the Tustin case we get

$$H(z) = \frac{0.417z^2 + 0.5z + 0.0830}{z^2 - 0.33z}$$

It should be mentioned that with all three of these methods some additional magnitude scaling may be needed if steady-state values are important. Also, although the trapezoidal method is generally considered the most effective approach, some distortion does occur at higher frequencies with this method, but it can be compensated for by a process called *prewarping*. In prewarping, s is changed to $\dfrac{\omega_o}{\omega_p} s$, where

$$\omega_p = \frac{2}{T}\tan\frac{\omega_0 T}{2} \tag{2.92}$$

and ω_0 is the frequency to be matched in the $H(s)|_{s=j\omega}$ and $H(z)|_{z=e^{j\omega T}}$ transfer functions. Bilinear transformation with frequency prewarping provides a good approximation to the analog transfer function. Example 4.2 shows the development of a digital equivalent of an analog transfer function using bilinear transformation with frequency prewarping.

Although there are several other methods of transforming an s-domain transfer function to an approximately equivalent z-domain form, such as the impulse invariance method or the differential mapping method, we conclude this chapter with a brief discussion of one more very popular method, the *matched pole-zero method*. In this method we map all the poles and zeros of the analog transfer function from the s-plane to the z-plane according to $z = e^{sT}$, where T is the sampling period. The following example illustrates the matched pole-zero method.

Example 2.34

Using the transfer function from Example 2.33,

$$H(s) = \frac{s+3}{(s+1)(s+2)}$$

we can write

$$s+3 \rightarrow \frac{1}{T}\ln z + 3$$

$$s+1 \rightarrow \frac{1}{T}\ln z + 1$$

$$s+2 \rightarrow \frac{1}{T}\ln z + 2$$

Setting the pole and zero terms to zero and solving for z, we get

Z = 0.0497

Z = 0.3678

Z = 0.135

so H(z) will be of the form

$$H(z) = \frac{k(z - 0.0497)}{(z - 0.3678)(z - 0.1350)}$$

and if we want H(s) at s = 0 to equal H(z) at z = 1, we get

$$\frac{3}{2} = \frac{k(0.95)}{(0.632)(0.865)}$$

Therefore, $k = 0.863$.

2.8 SUMMARY

The classical method of obtaining solutions to linear difference equations was covered in Section 2.2. The unit pulse response and discrete convolution were discussed in Section 2.3 along with ideas of zero-state and zero-input responses. The z-transform technique and some of its important properties such as linearity, convolution, and time shift were covered in Section 2.4 and the partial-fraction expansion method was introduced (in conjunction with Appendix D) as the basic method of inverse z-transformation. The concept of system transfer function and its cascaded, parallel, and canonical configurations was introduced in Section 2.5. In Section 2.6 we investigated the frequency response of a discrete system as well as ways to ferret out the frequency content of discrete signals using the Fourier transform. In particular, the DFT and IDFT were developed and discussed and the role played by the FFT in computing the discrete transforms was stressed. In Section 2.7 we presented the mapping relationship between s- and z-planes. Among other methods, the bilinear transformation method with frequency prewarping of an analog transfer function was discussed.

PROBLEMS

2.1 Find the complete solution for the following difference equations:

(a) $c(k) + c(k-1) + c(k-2) = 10^k$, $c(-1) = c(-2) = 0$

(b) $c(k-1) + 15c(k) = a^k$, $c(-1) = 2$

2.2 Construct a block-diagram representation for the following difference equation:

$$c(k) = 5r(k) + 2r(k-1) - 2c(k-1)$$

2.3 Find the difference equation representing the system shown in Fig. 2.25.

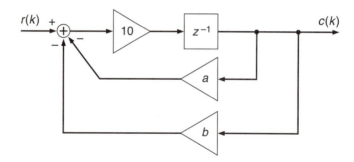

Figure 2.25   System for Problem 2.3.

2.4 Determine the z-transform of the following discrete signals

(a) $f(k) = ka^k u(k - 1)$

(b) $f(k) = ka^k (k - 1) u(k - 2)$

2.5 Use the linearity theorem to find the z-transform of the following signal:

$$f(k) = \delta(k) + 2a^{k-1} u(k)$$

2.6 Determine the initial value of a sequence that has the z-transform

$$F(z) = \frac{z}{z - a} \qquad\qquad \text{for } |z| > a$$

2.7 Determine the final value of a sequence that has the z-transform

$$F(z) = \frac{z^3}{z^3 - 1}$$

2.8 Use the time-shift property of the z-transform to find $z\{f(k + 1)\}$ and $z\{f(k - 1)\}$ for the following signals:

(a) $f(k) = 10k^2 u(k - 1)$

(b) $f(k) = (5k^2 - 10k)\, u(k + 1)$

(c) $f(k) = (k^2 - a^k)\, u(k - 2)$

2.9 Find the inverse z-transforms of the following functions:

(a) $F(z) = \dfrac{z^2}{(z-1)(z-2)}$

(b) $F(z) = \dfrac{z^2}{(z-1)^2(z-2)}$

2.10 Determine the response of the system governed by

$$c(k-2) - 0.5c(k-1) - 0.5c(k) = r(k)$$

to the input

$$r(k) = \delta(k) + \delta(k-1)$$

Assume zero initial conditions.

2.11 Given a system with input r(k) and output c(k) governed by

$$c(k) - c(k-1) = 3y(k) + y(k-1)$$

and

$$y(k) + y(k-1) = r(k)$$

find the equivalent cascaded system.

2.12 For the system of Problem 2.11, find the equivalent system using the parallel form.

2.13 For the system of Fig. 2.26, find the overall transfer function.

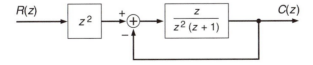

Figure 2.26 System for Problem 2.13.

2.14 For the system of Fig. 2.27, find the overall system transfer function and discuss
system stability.

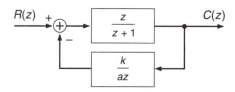

Figure 2.27   System for Problem 2.14.

2.15 Determine the poles and zeros of the following system transfer function and
discuss system stability:

$$H(z) = \frac{z(z+1)}{(z-1)(z+2)^2}$$

2.16 Let

$$H(z) = \frac{z^2 + z + 1}{z^4 + 0.25}$$

Determine and plot the unit step response and the unit pulse response. Use appropriate
MATLAB or MATRIX$_X$ tools.

2.17 Use the MATLAB statement RESIDUEZ to find the inverse z-transform of

$$F(z) = \frac{z^2 + 0.5z}{z^4 - 5z^2 + 4}$$

2.18 Determine and plot the magnitude and phase of the frequency response of a
system with a closed-loop transfer function

$$H(z) = \frac{z^5 + z^3 + z}{z^7 + z^6 + z^5 + z^2 + 1}$$

Use the MATLAB statement FREQZ.

2.19 For the system of Problem 2.18, construct the frequency response using the MATLAB statement BODE. Compare the results with the results of Problem 2.18.

2.20 Using MATLAB or MATRIX$_x$:

(a) Construct the Bode frequency response plot of the continuous-time system with

$$G(s) = \frac{s^2 + 0.5s + 0.5}{s^4 + 2s^3 + 3s^2 + 2s + 1}$$

Then convert G(s) to its G(z) equivalent using the bilinear transformation method with T=0.5.

(b) Construct the Bode plot of G(z) and compare it with G(s) plot. Repeat the comparison with T = 0.1 and T = 0.05.

2.21 Determine the DFT with N=8 if

(a)   f(n) = [1 0 0 0 0 0 0 0]
(b)   f(n) = [2 2 2 2 0 0 0 0]
(c)   f(n) = [1 2 0 0 1 2 0 0]
(d)   f(n) = [1 -1 1 -1 1 -1 1 -1]

2.22 Let f(k) = (-1)^{k}u(k) and N = 32. Determine the DFT using MATLAB and discuss your result. What kind of transform would you expect?

2.23 Find the closed-loop transfer function $C(z)/R(z)$ and the unit pulse response for the discrete system shown in Fig. 2.28 with

$$G_1(z) = \frac{z}{z - 0.5}$$

and

$$H_1(z) = \frac{z - 0.5}{z(z + 0.5)}$$

Is this system stable?

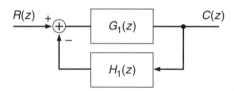

Figure 2.28   Block diagram for Problem 2.23.

2.24 A space telescope is represented in Fig. 2.29. A control is effected with proportional, integral, and rate feedback (this is the classic PID controller). Let $k_1 = k_2 = k_3 = 1$ and $k = 2$. Determine the closed-loop transfer function $C(s)/R(s)$. Then using the bilinear transformation, find the discrete version of $C(s)/R(s)$ assuming that $T = 0.1$.

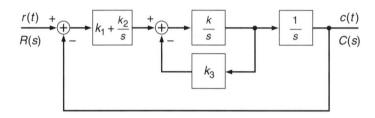

Figure 2.29 Block diagram for Problem 2.24.

REFERENCES

1.   C. R., Wylie, Jr., *Advanced Engineering Mathematics*, McGraw-Hill, New York, 1979.

2.   J. A, Cadzow, *Discrete-Time Systems,* Prentice Hall, Upper Saddle River, NJ, 1973.

3.   R. C., Dorf, *Modern Control Systems,* Addison-Wesley, Reading, MA, 1980.

4.   G. F. Franklin, and J. D. Powell, *Digital Control of Dynamic Systems,* Addison-Wesley, Reading, MA, 1980.

5.   H.F. Van Landingham, *Introduction to Digital Control Systems,* Macmillan New York, 1985.

6.   *Standard Mathematical Tables,* 25th ed., CRC Press, Boca Raton, FL, 1979.

7. C. Slivinsky and J. Borninski, *Control System Compensation and Implementation with the TMS32010. Digital Signal Processing Applications With the TMS320 Family: Theory, Algorithms, and Implementations,* Application Book Vol.1, Texas Instruments, Dallas, TX, 1989.

8. B.C. Kuo, *Digital Control Systems*, 2nd ed., Saunders College Publishing, Orlando, FL, 1992.

9. M. O'Flynn, and E. Moriarty, *Linear Systems: Time Domain and Transform Analysis*, J. Wiley, New York, 1987.

10. J.C. Hsu and A. Meyer, *Modern Control Principles and Applications,* McGraw-Hill, New York, 1968.

11. *The Student Edition of Matlab*, MathWorks, Inc., Prentice Hall, Upper Saddle River, NJ, 1992.

12. *Product Catalog,* DSP Technology Inc., Fremont, CA, 1990.

Analysis of Discrete Systems

3.1 INTRODUCTION

*I*n Chapter 1 we indicated that analysis is an important part of digital control system design methodology. In this chapter we introduce topics relevant to digital control system analysis*: sampled-data systems, state variables, nonlinear systems, stability*, and *sensitivity* analysis.

Sampled-data systems, which are systems that include both discrete and continuous signals, must be understood in order to model digitally controlled physical systems. Section 3.2 covers sampled-data systems as well as analog-to-digital, digital-to-analog, and resolver/synchro-to-digital converters. Section 3.2 also covers commonly used ADCs and DACs that interface to DSPs.

In Section 3.3 we discuss state-variable methods. These methods allow us to model systems based not only on their input–output relationships but also on their systems' internal states. These methods are attractive because we can use them with multiple input–output systems as well as nonlinear systems. In Section 3.3 we discuss state variables for both continuous and discrete systems. The reader might want to review Appendix E, which gives a brief overview of matrix algebra, before reading Section 3.3.

In Section 3.4 we discuss nonlinear discrete systems. Linearization techniques using Jacobian matrices are considered. Simulations of non-linear systems are considered employing the MATLAB/SIMULINK CAD tools. We conclude Section 3.4 with a brief look at chaos in nonlinear systems.

In Chapter 2 we described how the stable region of the s-plane maps into the z-plane. In Section 3.5 we extend this discussion by introducing the Jury stability method and the Liapunov technique. We also review the Routh–Hurwitz criterion and show how, with a few modifications, this familiar s-domain analog controls technique can be used in the z-domain.

In Section 3.6 we discuss the sensitivity of a system transfer function when a transfer function parameter changes.

We can use a control systems analyzer, as well as software packages such as $MATRIX_X$ and MATLAB not to only analyze control systems but to develop and test them as well. In Appendix A we discuss $MATRIX_X$ and MATLAB. Fig. 3.1 shows a control system analyzer that measures analog and digital signals and characterizes system stability with frequency response measurements.

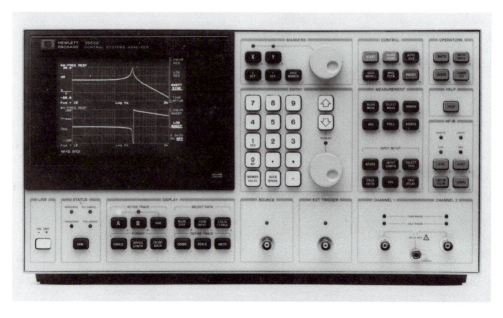

Figure 3.1   Control system analyzer (Photo Courtesy of the Hewlett-Packard Company.)
                   See Reference [9].

3.2 SAMPLED-DATA SYSTEMS

The overall model of a digitally controlled physical system contains both continuous and discrete signals. Systems with both types of signals are called *sampled-data systems.* Fig.

3.2 shows the basic block diagram of a sampled-data system. The digital computer is driven by discrete signals and the plant by continuous signals. The A/D converter converts the error signal e(t) to discrete form, or the error signal may already be of discrete form because the reference signal may be discrete and the feedback signal may need to be sent through an A/D converter. In any event, the signal that drives the digital computer is digitized. The computer will process the signal into it and produce a signal that will drive the plant such that the plant yields the desired output c(t). The output of the digital computer is of discrete form and needs to be converted to continuous form via a D/A converter.

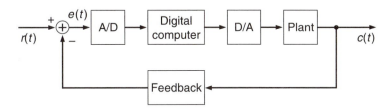

Figure 3.2   Basic block diagram of a sampled-data system.

Now, the simplest version of the sampled-data system of Fig. 3.2 is the case where the A/D converter, the digital computer, and the D/A converter are represented by a sample-and-hold device. Here, for example, the system output c(t) is the angle at which an antenna is pointed and the system input r(t) is the angle to an airplane the system is supposed to track automatically. Assuming unity gain feedback, the tracking error e(t) is r(t) - c(t). But if the radar transmits only once every T seconds, the error signal is only known at 0, T, 2T, 3T, ... A sampler models the conversion of the continuous signal to a discrete signal at discrete instants of time. We assume that an ideal sampler samples e(t) to yield e*(t) where

$$e * (t) = \sum_{k=0}^{\infty} e(kT)\delta(t - kT) \qquad (3.1)$$

(This equation was introduced in Chapter 2 and will be discussed further in this chapter.) Following the sampler is a data-hold device. The data-hold device reconstructs the discrete signal into a form resembling the signal before sampling.

The common method used to approximate the behavior of the data-hold device is to employ a Taylor series expansion of the error signal e(t). The expansion is

$$e(t) = e(nT) + e'(nT)(t - nT) + \frac{e''(nT)}{2}(t - nT)^2 + \cdots \qquad (3.2)$$

If we use only the first term of Equation 3.2 to approximate the hold device, then

$$e_n(t) = e(nT) \text{ for } nT < t < (n + 1)T \qquad (3.3)$$

This equation is the *zero-order hold* (ZOH). When we use the impulse modulation idea suggested in Chapter 2, the output of the ideal sampler is e*(t), as expressed in Equation 3.1. The Laplace transform of this is

$$E*(s) = \sum_{k=0}^{\infty} e(kT)e^{-skT} \tag{3.4}$$

Then, if the ZOH yields

$$e_n(t) = \sum_{k=0}^{\infty} e(kT)\big[u(t-kT) - u(t-kT-T)\big] \tag{3.5}$$

i.e., Equation 3.3 written as a step function approximation to e(t), then

$$E_n(s) = \sum_{k=0}^{\infty} e(kT)\left[\frac{e^{-skT}}{s} - \frac{e^{-s(kT+T)}}{s}\right] \tag{3.6}$$

Therefore, the transfer function of the ZOH is

$$\frac{E_n(s)}{E*(s)} = \frac{1}{s}[1 - e^{-sT}] \tag{3.7}$$

or

$$T(s)_{ZOH} = \frac{1}{s}[1 - e^{-sT}] \tag{3.8}$$

Since it fixes e(t) at a constant value for a T time interval, the zero-order-hold operation is sometimes called the *clamp* operation. If we use the first two terms of Equation 3.2, then

$$e_n(t) = e(nT) + e'(nT)(t-nT) \quad \text{for} \quad nT < t < (n+1)T \tag{3.9}$$

This equation is the *first-order hold* (FOH). The FOH has a transfer function

$$T(s)_{FOH} = (\frac{1}{s} + \frac{1}{s^2 T})[1 - e^{-sT}]^2 \tag{3.10}$$

which is quite complex compared to the ZOH. The second-order hold is even more complex and is seldom considered. For most applications, if the sampling is fast enough, the ZOH is quite adequate. It is by far the least expensive and most commonly used data-hold device.

Open-Loop Sampled-Data Structure

In the next two examples we consider the open-loop structure of the sampled data system involving only the sampler and the ZOH device in cascade with the plant transfer function.

Example 3.1

Fig. 3.3 shows a block diagram of a discrete-time system with a plant transfer function $G_p(s) = \frac{1}{s}$ In this system, signals are first sampled and then clamped before interfacing with the plant transfer function. The overall discrete-time transfer function of this system is

$$G(z) = Z\{(\frac{1 - e^{-sT}}{s})G_p\}$$

or

$$G(z) = (1 - z^{-1})Z\{\frac{G_p}{s}\}$$

where $G_p(s) = \frac{1}{s}$ Therefore,

$$G(z) = (1 - z^{-1})Z\frac{1}{s^2}$$

Of course, the z-transform of any function of s, G(s), really means the z-transform of the sampled version of the time function, which is the inverse Laplace transform of G(s).

$$G(z) = (1 - z^{-1})\{\frac{Tz}{(z-1)^2}\}$$

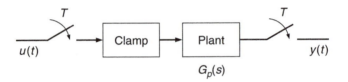

Figure 3.3   System for Example 3.1.

$$G(z) = \frac{T}{z-1}$$

There is an alternative approach to Example 3.1 using the "star" notation introduced in Chapter 2. Let the Laplace transform of the outputs of the samplers be U*(s) and Y*(s) respectively. Then

$$Y(s) = U*(s)T(s)_{ZOH} G_p(s)$$

$$= U*(s)[\frac{1-e^{-sT}}{s^2}]$$

Then

$$Y*(s) = U*(s)[\frac{1-e^{-sT}}{s^2}]*$$

and since

$$X*(s) = X(z) \text{ with } e^{sT} = z$$

we get

$$Y(z) = U(z)(1-z^{-1})Z\{\frac{1}{s^2}\}$$

$$= U(z)(1-z^{-1})\frac{Tz}{(z-1)^2}$$

$$= U(z)\frac{T}{z-1}$$

from which

$$G(z) = \frac{Y(z)}{U(z)} = \frac{T}{z-1}$$

as before.

The mathematics involved in these determinations of sampled transfer functions can get quite involved, especially if the order of the plant is more than just one or two. Fortunately, there are CAE tools available to make the required determinations. The DISCRETIZ statement from MATRIX$_X$ provides a convenient way to obtain G(z) as

$$G(z) = (1 - z^{-1})Z\left\{\frac{G_p(s)}{s}\right\} \tag{3.11}$$

The next example illustrates its use.

Example 3.2

Assume that a plant has a transfer function

$$G_p(s) = \frac{8s(s+2)}{s^3 + 9s^2 + 23s + 15}$$

Convert the system into its clamped z-transformed expression G(z).

Solution: Let N=[8 16 0] and D=[1 9 23 15] and assume a sample period DT=1.0. Then <ND,DD> = DISCRETIZ (N,D,DT) returns

DD=[1 -0.424 0.0211 -0.0001] and ND=[0.4472 -0.4969 0.0497]

which, implies a transfer function of the form

$$G(z) = \frac{0.4472z^2 - 0.4969z + 0.0497}{z^3 - 0.424z^2 + 0.0211z - 0.0001}$$

As expected, since the poles of $G_p(s)$ are at $s = -1, -3, -5$, the denominator of $G(z)$ factors into $(z - e^{-T})(z - e^{-3T})(z - e^{-5T})$, where $T = 1.0$. Also, the numerator factors into $(z - 1)(0.4472z - 0.0497)$ containing the numerator of the $(1 - z^{-1})$ term.

The preceding type of analysis, employing the ideal sampler and the ZOH, is common to most sample-and-hold operations and can be used to approximate even very sophisticated DAC and ADC procedures. We look next at some of the hardware involved in the architecture of sample-and-hold devices, DACs, and ADCs.

Digital-to-Analog Converters

Two categories of digital-to-analog converters (DACs) exist: general-purpose and function-specific. General-purpose DACs perform the basic data conversion from the digital domain to the analog domain. Fig. 3.4 shows the basic architecture of a simple digital-to-analog converter. The architecture consists of a ladder network of resistors, an AND array for switching, and an op-amp circuit for output scaling.

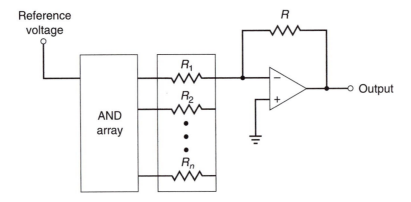

Figure 3.4   Architecture of a simple general-purpose DAC.

Function-specific DACs have optimized architectures that make them suitable for specific applications such as video RAM and audio DAC. Some commonly used DACs interface directly to a microprocessor or a DSP. Fig. 3.5 shows the basic block diagram of a DAC that interfaces to DSPs.

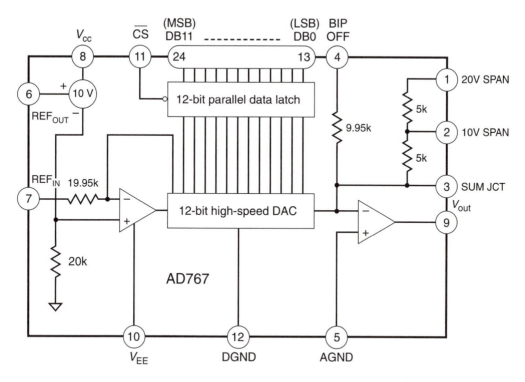

Figure 3.5   Functional block diagram of a DAC designed to interface to a DSP. (Courtesy of Analog Devices, Inc.) See References [6,7]

The AD767 is a 12-bit DAC that uses an on-chip latch to make it compatible with a DSP. The 12-bit single buffered-input register accepts 12-bit parallel data from DSPs such as the TMS320 series devices, the ADSP-2100, and microprocessors such as the 68000 and 8086. In Fig. 3.6 we present a typical DSP/DAC interface.

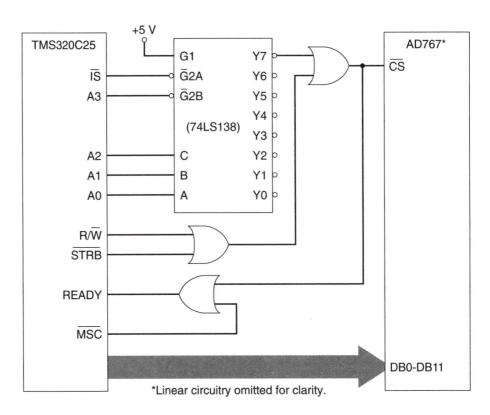

Figure 3.6   TMS320C25–AD767 interface. (Courtesy of Analog Devices, Inc.) See
            References [6,7]

The nature of the digital input to a DAC depends on factors such as coding (binary, BCD, two's complement, and so on) and type of logic (positive or negative). There are many criteria for selecting a DAC, the most important of which are:

• *Dynamic range:* the ratio of the largest output to the smallest output
• *Resolution:* the number of binary words
• *Switching time:* the time the switch uses to change from one state to the other

Analog-to-Digital Converters

Fig. 3.7 shows the basic architecture of a simple general-purpose analog-to-digital converter (ADC).

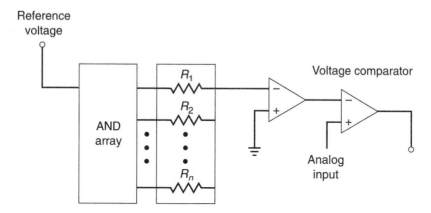

Figure 3.7   Basic architecture of a simple general-purpose ADC.

This architecture contains a single comparator which compares the analog input with the output of an n-bit DAC. The following criteria are used to select an ADC:

• *Resolution:* the number of states into which the analog input voltage is divided

• *Conversion time:* the time the ADC requires for a complete measurement

• *Cycle time:* the time required for the converter clock to finish a conversion

• *Analog input:* bipolar or unipolar

• *Bandwidth*

• *Whether or not the input signal is already sampled*

• *Sample-and-hold capability*

An ADC requires that its sample-and-hold function maintain a constant input value during conversion. If the sample-and-hold device is external to the ADC, it must precede the ADC. Fig. 3.8 shows a sample-and-hold device that is external to its ADC. The time required to obtain the sample signal and the time required to go from sample to hold are two important variables that determine its sample-and-hold capability.

Fig. 3.9 shows the functional block diagram of the AD7878, which is an ADC designed to interface to a DSP. The main parts of its architecture are a built-in FIFO (first-in, first-out) memory and control logic. This ADC is compatible with all 16-bit microprocessors and 16-bit DSPs. We can use the AD7878 in DSP servo control, high-speed modems, and for DSP applications. Fig. 3.10 shows the interface between the AD7878 ADC and the TMS320 DSP.

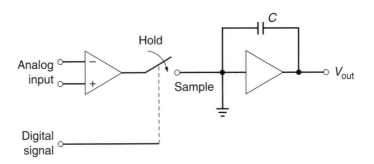

Figure 3.8   Sample-and-hold device.

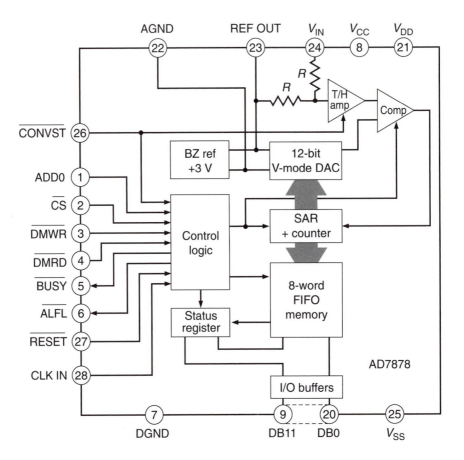

Figure 3.9   Functional block diagram of an ADC designed to interface to a DSP. (Courtesy of Analog Devices, Inc.) See References [6,7]

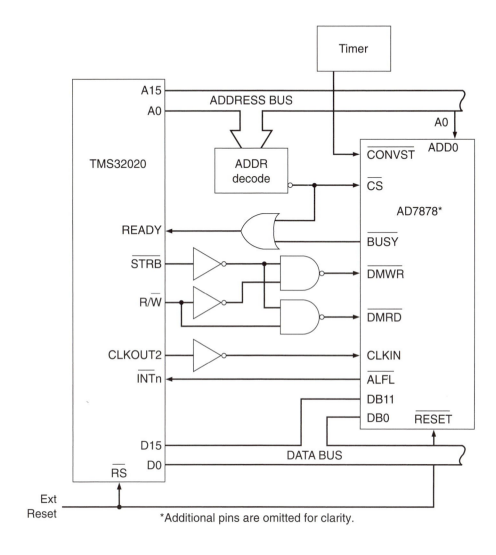

Figure 3.10   AD7878–TMS32020 interface. (Courtesy of Analog Devices, Inc.) See
References [6,7]

Resolver/Synchro-to-Digital Converters

Resolver/synchro-to-digital converters are widely used in control systems for digital measurement and for control of linear and angular displacements. Table 3.1 lists some of their applications.

Table 3.1   Applications for Resolver/Synchro-to-Digital Converters

Angular Displacement	Linear Displacement
Gimbal and gyro control systems	Industrial gauging
Engine controllers	Linear positioning systems
Servo control systems	Linear actuator control
Antenna monitoring	Factory automation
Robotics	Industrial process control
Stabilization systems	Automotive motion sensing and control

Resolver/synchro-to-digital converters have either a four-wire resolver format (sin A, cos A, sin B, cos B) or a three-wire synchro format, as well as two reference inputs (Ref. A and Ref. B). The output of a resolver/synchro-to-digital converter is a digital word representing the shaft angle of the transducer.

Closed-Loop Sampled-Data Structure

Having considered an open-loop sampled-data structure and some of the details of ADC and DAC conversion, let us now look at Fig. 3.2 again and model the DAC with a ZOH transfer function, $T(s)_{ZOH} = T(s)$, the digital computer with a transfer function, $D(z) = D^*(s)$, and the ADC converter as an ideal sampler. Assume that the feedback is unity gain and we can represent the entire system as in Fig. 3.11.

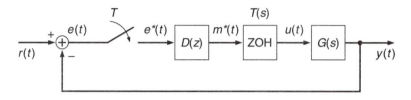

Figure 3.11   Another block diagram of a sampled-data system.

From this block diagram we can write

$$e(t) = r(t) - y(t) \tag{3.12}$$

and taking the Laplace transform, we get

$$E(s) = R(s) - Y(s). \tag{3.13}$$

Then

$$E^*(s) = R^*(s) - Y^*(s) \tag{3.14}$$

and if the output of the D(z) block is $M^*(s)$ then

$$M^* = E^*D^* \tag{3.15}$$

dropping the s argument. The output of the ZOH, U(s), can be written as

$$U = M^*T \tag{3.16}$$

or

$$U^* = M^*T^* \tag{3.17}$$

Then we can write

$$Y = GU = GM^*T \tag{3.18}$$

and

$$
\begin{aligned}
Y^* &= (GT)^* M^* \\
&= (GT)^*(R^* - Y^*)D^* \\
&= (GT)^*D^*R^* - (GT)^*D^*Y^*
\end{aligned}
\tag{3.19}
$$

from which we can solve for the transfer function

$$\frac{Y^*}{R^*} = \frac{(GT)^*D^*}{1 + (GT)^*D^*} \tag{3.20}$$

or

$$T(z) = \frac{Y(z)}{R(z)} = \frac{GT(z)D(z)}{1 + GT(z)D(z)} \qquad (3.21)$$

which is the familiar $G/(1 + G)$ form for unity gain feedback closed-loop transfer functions. Note, however, that the G(s) and T(s) transfer functions must be handled as a pair. A major part of our concern in the rest of the book will be to design a D(z) transfer function given a plant transfer function G(s) and using the ZOH transfer function T(s). Once a suitable D(z) is determined, we will consider its implementation with DSPs. We discuss conventional D(z) design in Chapter 4 and DSPs as controllers in Chapter 5. Then in Chapter 6 we consider modern control methods using DSPs. At this stage, we can do some minimal design manipulations, as illustrated in the following example.

Example 3.3

Let GT(z) be the result we got from Example 3.1 with T = 1, namely, $1/(z - 1)$. Assume that we want a closed-loop transfer function $T(z) = z/(z^2 - 0.25)$. Then algebraically we can solve for the required D(z):

$$D(z) = \frac{(z-1)z}{(z^2 - z - 0.25)}$$

Note that all the transfer functions in a sampled-data system, whether s- or z-domain functions, are frequency-domain descriptions. As indicated in Chapter 2, time-domain descriptions are also valuable for discrete and continuous systems. Of special importance are state-variable time-domain representations of dynamic systems, whether continuous- or discrete-time systems, or combinations of continuous- and discrete-time systems as in sampled-data systems. State-variable representations are fundamental to most modern control developments. We take them up next.

3.3 STATE-VARIABLE METHODS

The classical method for understanding a system is to analyze the relationships between the sets of input and output variables. An example of this method is transfer function analysis. In transfer function analysis we consider the system to be a black box. System formulation is based on the cause (system inputs) and effect (system outputs) notion. The ratio of output to input Laplace or z-transforms gives us a system description. The state-variable method, on the other hand, takes into account the internal states of a system as well as its inputs and outputs (Fig. 3.12).

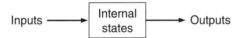

Figure 3.12   Linear system and its internal and external variables.

In referencing both internal and external variables, the state-variable method more fully describes a system and provides generally a more powerful system formulation. But the mathematics involved is more complex. Unlike the transfer function approach, which uses simple ratios of polynomials, matrix notation is employed in the state-variable formulation. What we gain, however, is that the matrix notation of the state-variable methods allows us to handle a wider class of systems. State-variable methods apply to both linear and nonlinear systems and can handle nonzero initial conditions as well as multiple input–output systems. These methods are widely used with commercial CAE packages.

Definition 3.1. The *state* of a system is the minimum set of internal variables of a system, which together with the input r(t), uniquely determine the output c(t).

Example 3.4

As an example of a system's state, consider a simple object in four-dimensional space-time. Under the right conditions, if we apply a force F, the object will displace in some way (Fig. 3.13).

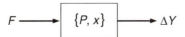

Figure 3.13   Object moving in space-time from application of a force.

In this example, force F is the input, the amount of displacement ΔY is the output, and momentum P (P = mv) and position, x, together can be considered as the internal states of the system.

Example 3.5

Selecting state variables is usually based on isolating the key physical variables within a system. For electrical circuits a consideration of energy storage devices is essential. Fig. 3.14 shows a simple linear RLC network. In this network, the inductor current i(t) associated with energy stored in the inductor $\left[W_L = \frac{1}{2}(i(t)^2 L)\right]$ and capacitor voltage v(t) associated with the energy stored in the capacitor $\left[W_C = \frac{1}{2}(i(t)^2 C)\right]$ are selected as state variables.

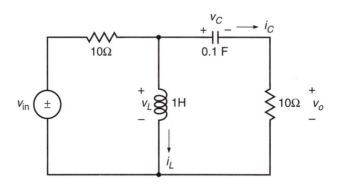

Figure 3.14   Linear RLC network.

We can write node equations at the top-right and top-middle nodes:

$$(v_{in} - v_L)0.1 = i_L + i_C \text{ and } i_C = 0.1v_O$$

But $i_C = 0.1(\frac{dv_c}{dt})$ and $V_L = \frac{dv_c}{dt}$
Therefore,

$$V_{in} = V_L + 10i_L + 10i_C = \frac{di_L}{dt} + 10i_L + \frac{dv_C}{dt}$$

Also, a mesh equation around the right-hand mesh yields $-v_L + v_C + 10i_C = 0$ or

$$\frac{di_L}{dt} = v_C + 10i_C = v_C + \frac{dv_C}{dt}$$

Then

$$V_{in} = v_c + \frac{dv_c}{dt} + 10i_L + \frac{dv_c}{dt} \text{ or } \frac{dv_c}{dt} = 0.5v_{in} - 0.5v_C - 5i_L$$

and

$$\frac{di_L}{dt} = v_{in} - 10i_L - 0.5v_{in} + 0.5v_C + 5i_L = 0.5v_{in} - 5i_L + 0.5v_C$$

Then we can arrange the derivative terms on the left and put these equations into matrix form as follows:

$$\frac{d}{dt}\begin{bmatrix} i_L \\ v_c \end{bmatrix} = \begin{bmatrix} -5 & 0.5 \\ -5 & -0.5 \end{bmatrix}\begin{bmatrix} i_L \\ v_c \end{bmatrix} + \begin{bmatrix} 0.5 \\ 0.5 \end{bmatrix} v_{in}$$

This is a standard state-variable format involving a state vector $\begin{bmatrix} i_L \\ v_C \end{bmatrix}$, the input v_{in}, and a 2×2 system matrix.

Because we use both continuous and discrete representations to study sampled-data systems, we must formulate state-variable descriptions of both continuous and discrete systems.

State-Variable Description of Continuous Systems

After defining state-variables of a system, we can model the state-variable description of a continuous-time system from the corresponding differential equations of the system, like the node and mesh equations used to describe the circuit of Example 3.5. Fig. 3.15 shows a continuous-time system where the state, the input, and the output are generally vectors of different dimensions. Equations 3.22 and 3.23 formulate the state-variable description of this system.

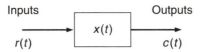

Figure 3.15 Continuous-time system.

$$\frac{dx(t)}{dt} = Ax(t) + Br(t) \tag{3.22}$$

and

$$c(t) = Cx(t) + Dr(t) \tag{3.23}$$

where x(t), the state vector, is an n-dimensional vector; r(t), the input vector, is an m-dimensional vector; c(t), the output vector, is a k-dimensional vector; A is an $n \times n$ matrix; B is an $n \times m$ matrix; C is a $k \times n$ matrix; and D is a $k \times m$ matrix. Vector notation allows us to use state variables for both single and multiple input-output descriptions.

Equation 3.22 (the state equation) describes the rate of change of the state x(t) as a function of the state x(t) and input r(t). Equation 3.23 (the output equation) describes the output c(t) as a function of the state x(t) and input r(t). Equations 3.22 and 3.23 are an alternative formulation to the transfer function method, which describes the system under study only in reference to its input $r(t) \Leftrightarrow R(s)$ and its output $c(t) \Leftrightarrow C(s)$. Equations 3.22 and 3.23 are called the dynamical equations of a continuous-time system.

Example 3.6

Consider the field-controlled DC motor system shown Fig. 3.16, where $\theta =$ Angular displacement; k = proportional constant; $I_a =$ armature current; $I_f =$ field current; J = inertia of the load; and f = friction.

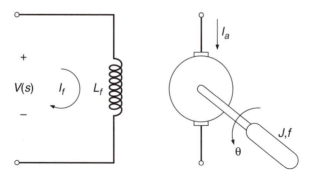

Figure 3.16   Field-controlled DC motor system.

The transfer function description of this system is

$$T(s) = \frac{\theta(s)}{v(s)} = \frac{k}{s(sJ + f)\ (Ls + R)}$$

We can obtain the state-variable representation of the system from the differential equations that describe the electrical and mechanical subsystems. These equations are

$$L\frac{di(t)}{dt} + Ri(t) = v(t)$$

and

$$J\frac{d^2\theta(t)}{dt^2}+f\frac{d\theta(t)}{dt}=ki(t)$$

The first equation is from Kirchhoff's voltage law (KVL) for the electrical subsystem. The second equation is for the mechanical subsystem. The states of the system are

$$x_1 = i$$
$$x_2 = \theta$$
$$x_3 = \dot{\theta}$$

and manipulating the two differential equations, we can write

$$\frac{dx_1}{dt} = -\frac{R}{L}x_1 + \frac{1}{L}v(t)$$

$$\frac{dx_2}{dt} = x_3$$

$$\frac{dx_3}{dt} = -\frac{f}{J}x_3 + \frac{k}{J}x_1$$

These can be put into the vector/matrix state-variable form as follows:

$$\begin{bmatrix}\dot{x}_1 \\ \dot{x}_2 \\ \dot{x}_3\end{bmatrix} = \begin{bmatrix}\dfrac{-R}{L} & 0 & 0 \\ 0 & 0 & 1 \\ \dfrac{K}{J} & 0 & \dfrac{-f}{J}\end{bmatrix}\begin{bmatrix}x_1 \\ x_2 \\ x_3\end{bmatrix} + \begin{bmatrix}\dfrac{1}{L} \\ 0 \\ 0\end{bmatrix}v(t)$$

and

$$\theta(t) = \begin{bmatrix}0 & 1 & 0\end{bmatrix}x(t)$$

Now, although multiple-input/multiple-output (MIMO) systems can be represented in the state-variable format, in subsequent discussions, to avoid unwieldy formulations we restrict ourselves for the most part to single-input/single-output (SISO) systems. For such systems we normally have available a scalar transfer function description H(s). Let H(s) take the general form

$$H(s) = \frac{c_{n-1}s^{n-1} + \cdots + c_1 s + c_0}{s^n + a_{n-1}s^{n-1} + \cdots + a_1 s + a_0} + b_n \tag{3.24}$$

We can transform a given H(s) representation into several special or *canonical form* state-variable realizations. These are special for various reasons, generally because they possess interesting mathematical properties. We look at three of the most important of these canonical forms: (1) the controllability form (CF), (2) the observability form (OF), and (3) the Jordan form (JF).

The CF and OF realizations are available by reading coefficients directly from H(s). Write Equations 3.22 and 3.23 in the form

$$\frac{dx_c(t)}{dt} = A_c x_c(t) + B_c r(t) \tag{3.25}$$

$$c(t) = C_c x_c(t) + D_c r(t) \tag{3.26}$$

where

$$A_c = \begin{bmatrix} 0 & 1 & 0 & \cdots & 0 \\ 0 & 0 & 1 & \cdots & 0 \\ \vdots & \vdots & \vdots & \vdots & \vdots \\ 0 & 0 & 0 & \cdots & 1 \\ -a_0 & -a_1 & -a_2 & \cdots & -a_{n-1} \end{bmatrix} \tag{3.27}$$

$$B_c = \begin{bmatrix} 0 \\ 0 \\ \vdots \\ 0 \\ 1 \end{bmatrix} \tag{3.28}$$

$$C = \begin{bmatrix} c_0 & c_1 & c_2 & \cdots & c_{n-1} \end{bmatrix} \qquad (3.29)$$

and

$$D_C = \begin{bmatrix} b_n \end{bmatrix} \qquad (3.30)$$

We can also write Equations 3.22 and 3.23 in the form

$$\frac{dx_o(t)}{dt} = A_o x_o(t) + B_o r(t) \qquad (3.31)$$

$$c(t) = C_o x_o(t) + D_o r(t) \qquad (3.32)$$

where

$$A_o = \begin{bmatrix} 0 & 0 & \cdots & 0 & -a_0 \\ 1 & 0 & \cdots & 0 & -a_1 \\ 0 & 1 & \cdots & 0 & -a_2 \\ \vdots & \vdots & & \vdots & \vdots \\ 0 & 0 & \cdots & 1 & -a_{n-1} \end{bmatrix} \qquad (3.33)$$

$$B_o = \begin{bmatrix} c_0 \\ c_1 \\ c_2 \\ \vdots \\ c_{n-1} \end{bmatrix} \qquad (3.34)$$

$$C_O = \begin{bmatrix} 0 & 0 & \cdots & 0 & 1 \end{bmatrix} \qquad (3.35)$$

and

$$D_O = \begin{bmatrix} b_n \end{bmatrix} \qquad (3.36)$$

Note that

$$A_C = A_O^T \qquad (3.37)$$

$$B_C = C_O^T \qquad (3.38)$$

$$C_C = B_O^T \qquad (3.39)$$

$$D_c = D_o \qquad (3.40)$$

Example 3.7

Cast the system with transfer function

$$H(s) = \frac{s^3 + s^2 + 2s + 2}{s^3 + 2s^2 + 3s + 3}$$

into the controllability canonical form.

Solution: Note that the system is not quite in the form indicated by Equation 3.24. We must first perform long division to write it in the form

$$H(s) = 1 + \frac{-s^2 - s - 1}{s^3 + 2s^2 + 3s + 3}$$

Then we can write

$$a_0 = 3$$

$$a_1 = 3$$

$$a_2 = 2$$

$$c_0 = -1$$

$$c_1 = -1$$

$$c_2 = -1$$

$$b_n = b_3 = 1$$

from which it follows that

$$A_C = \begin{bmatrix} 0 & 1 & 0 \\ 0 & 0 & 1 \\ -3 & -3 & -2 \end{bmatrix}$$

$$B_C = \begin{bmatrix} 0 \\ 0 \\ 1 \end{bmatrix}$$

and

$$D_C = \begin{bmatrix} 1 \end{bmatrix}$$

The n-dimensional state vectors in the CF and OF formulations, $x_C(t)$ and $x_O(t)$ respectively, are not usually real physical variables internal to the system. But the CF and OF realizations do exhibit some useful mathematical properties that we investigate later. Also, given only the transfer function H(s), we generally do not know what the internal physical structure of the system is anyway. However, if we do know the internal structure of the system as we did in Examples 3.5 and 3.6, the state-variable expression in the real physical variables might be the preferred representation.

In any event, given a physical variable state-space form or a CF or OF form, we can transform any of these into another useful canonical form, the Jordan canonical form (JF). The JF has the useful property that its matrix is diagonal (or almost diagonal). That means the states are decoupled and the solution of the state equations is generally simplified. The next example illustrates this.

Example 3.8

Let

$$\frac{dx}{dt} = \begin{bmatrix} -2 & 0 & 0 \\ 0 & -5 & 0 \\ 0 & 0 & -10 \end{bmatrix} x + \begin{bmatrix} 1 \\ 2 \\ 1 \end{bmatrix} r$$

and assume that the input r(t) is zero and the initial condition vector at time t = 0 is

$$x(0) = \begin{bmatrix} 2 \\ 1 \\ -1 \end{bmatrix}$$

The state equations become

$$\frac{dx_1}{dt} = -2x_1$$

$$\frac{dx_2}{dt} = -5x_2$$

$$\frac{dx_3}{dt} = -10x_3$$

Therefore,

$$x_3(t) = K_3 e^{-10t} \text{ and at } t = 0 \text{ we get } x_3(0) = -1 = K_3$$

$$x_2(t) = K_2 e^{-5t} \text{ and } x_2(0) = K_2 = 1$$

$$x_1(t) = K_1 e^{-2t} \text{ and } x_1(0) = K_1 = 2$$

Thus the solution of the state equation becomes

$$x(t) = \begin{bmatrix} 2e^{-2t} \\ e^{-5t} \\ -e^{-10t} \end{bmatrix} \quad t \geq 0$$

The problem is: How do we cast a given system into the JF? Any state vector $x(t)$ can be transformed into another state vector $\hat{x}(t)$ by means of a coordinate transformation P such that

$$\hat{x}(t) = Px(t) \qquad (3.41)$$

where P is an $n \times n$ nonsingular constant matrix. We now present a method of determining P when $\hat{x}(t)$ is given and $x(t)$ is the state vector of the JF. We will assume that the A matrix of the JF is diagonal. The terms on the diagonal of the A matrix are called *eigenvalues* of the system. They are actually the roots of the system characteristic equation, the poles of the system transfer function. (We consider only the case where the roots or eigenvalues are nonrepeated, because if the roots are repeated, there can be "1"s in the off-diagonal of the A matrix and the determinations get more involved. We recommend that CAE programs be used to analyze these more complicated situations.) Eigenvalues, λ, can be obtained by solving the equation

$$\det[\lambda I - \hat{A}] = 0 \qquad (3.42)$$

We assume that the given system is described by $\hat{A}, \hat{B}, \hat{C}, \hat{D}$, that is,

$$\frac{d\hat{x}}{dt} = \hat{A}\hat{x} + \hat{B}r \qquad (3.43)$$

and

$$y = \hat{C}\hat{x} + \hat{D}r \qquad (3.44)$$

Letting $\hat{x}(t) = Px(t)$, we get

$$\frac{d\hat{x}}{dt} = \hat{A}Px + \hat{B}r = P\frac{dx}{dt} \qquad (3.45)$$

and

$$y = \hat{C}Px + \hat{D}r \tag{3.46}$$

Thus

$$\frac{dx}{dt} = (P^{-1}\hat{A}P)x + (P^{-1}\hat{B})r \tag{3.47}$$

and the new JF system matrices are

$$\begin{aligned} A &= P^{-1}\hat{A}P \\ B &= P^{-1}\hat{B} \\ C &= \hat{C}P \\ D &= \hat{D} \end{aligned} \tag{3.48}$$

To determine P let

$$P = \begin{bmatrix} p_1 & \vdots & p_2 & \vdots & \cdots & \vdots & p_n \end{bmatrix} \tag{3.49}$$

where the p_i are called *eigenvectors* of A and can be determined from

$$[\lambda_i I - \hat{A}]p_i = 0 \text{ for } i = 1, 2, ..., n \tag{3.50}$$

The following example illustrates the procedure for diagonalizing a state-variable representation.

Example 3.9

Given the system

$$\hat{A} = \begin{bmatrix} -1 & 0 & 0 \\ 2 & -2 & 0 \\ 1 & 1 & -3 \end{bmatrix}$$

$$\hat{B} = \begin{bmatrix} 1 \\ 2 \\ -1 \end{bmatrix}$$

$$\hat{C} = \begin{bmatrix} 0 & 1 & 0 \end{bmatrix}$$

$$\hat{D} = [0]$$

convert it to the Jordan canonical form.

Solution: The eigenvalues are readily obtained from $\det[\lambda I - \hat{A}] = 0$:

$$\lambda_1 = -1$$
$$\lambda_2 = -2$$
$$\lambda_3 = -3$$

Then for $\lambda_1 = -1$, $[\lambda_1 I - \hat{A}]p_1 = 0$ yields

$$\left(\begin{bmatrix} -1 & 0 & 0 \\ 0 & -1 & 0 \\ 0 & 0 & -1 \end{bmatrix} - \begin{bmatrix} -1 & 0 & 0 \\ 2 & -2 & 0 \\ 1 & 1 & -3 \end{bmatrix} \right) \begin{bmatrix} p_{11} \\ p_{21} \\ p_{31} \end{bmatrix} = \begin{bmatrix} 0 \\ 0 \\ 0 \end{bmatrix}$$

and we get 0=0, $-2p_{11} + p_{21} = 0$, $-p_{11} - p_{21} + 2p_{31} = 0$. One of the interesting things about this procedure is that we get to choose one of the P values arbitrarily. This is because we have only two independent equations in three unknowns. The simpler the choice the better, so let $p_{11} = 1$, then $p_{21} = 2$ and $p_{31} = 1.5$. Thus

$$p_1 = \begin{bmatrix} 1 \\ 2 \\ 1.5 \end{bmatrix}$$

The same procedure for λ_1 and λ_2 yields

$$p_2 = \begin{bmatrix} 0 \\ 1 \\ 1 \end{bmatrix} \text{ and } p_3 = \begin{bmatrix} 0 \\ 0 \\ 1 \end{bmatrix}$$

Therefore,

$$P = \begin{bmatrix} 1 & 0 & 0 \\ 2 & 1 & 0 \\ 1.5 & 1 & 1 \end{bmatrix} \text{ and } P^{-1} = \begin{bmatrix} 1 & 0 & 0 \\ -2 & 1 & 0 \\ 0.5 & -1 & 1 \end{bmatrix}$$

and we get

$$A = P^{-1} \hat{A} P = \begin{bmatrix} 1 & 0 & 0 \\ -2 & 1 & 0 \\ 0.5 & -1 & 1 \end{bmatrix} \begin{bmatrix} -1 & 0 & 0 \\ 2 & -2 & 0 \\ 1 & 1 & -3 \end{bmatrix} \begin{bmatrix} 1 & 0 & 0 \\ 2 & 1 & 0 \\ 1.5 & 1 & 1 \end{bmatrix} = \begin{bmatrix} -1 & 0 & 0 \\ 0 & -2 & 0 \\ 0 & 0 & -3 \end{bmatrix}$$

Also,

$$B = P^{-1} \hat{B} = \begin{bmatrix} 1 \\ 0 \\ -1.5 \end{bmatrix} \quad C = \hat{C} P = \begin{bmatrix} 2 & 1 & 0 \end{bmatrix} \quad D = \hat{D} = [0]$$

There is a MATRIX$_x$ statement <D,V>=EIG(A) that will take any $n \times n$ A matrix as input and return its diagonal form D and also V, which is our matrix of eigenvectors, the P matrix. Using Example 3.9 with $\hat{A}$ as input we get

$$A = D = \begin{bmatrix} -1 & 0 & 0 \\ 0 & -2 & 0 \\ 0 & 0 & -3 \end{bmatrix} \text{ but } V = P = \begin{bmatrix} 0.3714 & 0 & 0 \\ 0.7428 & 0.7071 & 0 \\ 0.5571 & 0.7071 & 1 \end{bmatrix}$$

which is not the same as our P matrix. This only illustrates the nonuniqueness of the eignvectors that constitute the P matrix.

Any given (A, B, C, D) system has infinitely many other representations several of which are valuable. The CF representation has the advantage of being *controllable* and the

OF has the advantage of being *observable.* Controllability and observability are important features of a system and are considered in Chapter 6 when we look at modern design techniques. The JF representation has the advantage of easy-to-solve state equations. As we have seen, we can get the JF from a given (A, B, C, D) representation in a straightforward but sometimes tedious fashion. But the OF and CF are most easily determined from the transfer function of a given system. Given an (A, B, C, D) representation we can construct the system transfer function quite easily by taking the Laplace transform of Equations 3.22 and 3.23 assuming zero initial conditions:

$$sX(s) - 0 = AX(s) + BR(s) \tag{3.51}$$

and

$$C(s) = CX(s) + DR(s) \tag{3.52}$$

Then solving Equation 3.51 for X(s) and substituting it into Equation 3.52, we get

$$C(s) = C(sI - A)^{-1}BR(s) + DR(s) = [C(sI - A)^{-1}B + D]R(s) \tag{3.53}$$

and for SISO systems we have the scalar transfer function

$$H(s) = \frac{C(s)}{R(s)} = C(sI - A)^{-1}B + D \tag{3.54}$$

Example 3.10

Consider the system of Example 3.9 with

$$A = \begin{bmatrix} -1 & 0 & 0 \\ 2 & -2 & 0 \\ 1 & 1 & -3 \end{bmatrix}$$

$$B = \begin{bmatrix} 1 \\ 2 \\ -1 \end{bmatrix}$$

$$C = \begin{bmatrix} 0 & 1 & 0 \end{bmatrix}$$

$$D = \begin{bmatrix} 0 \end{bmatrix}$$

We saw how to cast this system into the JF. Now consider the CF and OF representations.
First construct the system transfer function

$$H(s) = \begin{bmatrix} 0 & 1 & 0 \end{bmatrix} \begin{bmatrix} s+1 & 0 & 0 \\ -2 & s+2 & 0 \\ -1 & -1 & s+3 \end{bmatrix}^{-1} \begin{bmatrix} 1 \\ 2 \\ -1 \end{bmatrix} + \begin{bmatrix} 0 \end{bmatrix}$$

or

$$H(s) = \frac{2(s+3) + 2(s+1)(s+3)}{(s+1)(s+2)(s+3)} = \frac{2s^2 + 10s + 12}{s^3 + 6s^2 + 11s + 6}$$

Note that we can cancel the $(s+3)$ term and in fact the $(s+2)$ as well. But we will not
effect these cancellations and will discuss this type of problem in Chapter 6.   The
coefficients are readily available from the transfer function:

$c_2 = 2$

$c_1 = 10$

$c_0 = 12$

$a_0 = 6$

$a_1 = 11$

$a_2 = 6$

and from these we form the matrices

$$A_C = \begin{bmatrix} 0 & 1 & 0 \\ 0 & 0 & 1 \\ -6 & -11 & -6 \end{bmatrix} \quad B_C = \begin{bmatrix} 0 \\ 0 \\ 1 \end{bmatrix} \quad C_C = \begin{bmatrix} 12 & 10 & 2 \end{bmatrix} \quad D_C = \begin{bmatrix} 0 \end{bmatrix}$$

$$A_o = \begin{bmatrix} 0 & 0 & -6 \\ 1 & 0 & -11 \\ 0 & 1 & -6 \end{bmatrix} \quad B_o = \begin{bmatrix} 12 \\ 10 \\ 2 \end{bmatrix} \quad C_o = \begin{bmatrix} 0 & 0 & 1 \end{bmatrix} \quad D_C = \begin{bmatrix} 0 \end{bmatrix}$$

There are several MATRIX$_x$ and MATLAB expressions that ease the burden of some of these computations. Considering just MATLAB, for instance, TF2SS will take a transfer function and convert it to a state-space representation, and SS2TF will take an A, B, C, D representation and convert it to its transfer function representation.

Example 3.11

Using the results of Example 3.10, take the CF matrices and reconstruct the transfer function using the SS2TF statement from MATLAB. Let

> A = [0 1 0; 0 0 1; -6 –11 –6]
> B = [0 0 1]′ (the prime implies transpose)
> C = [12 10 2]
> D = [0]

then the MATLAB statement

> [NUM, DEN] = SS2TF (A, B, C, D, 1)

returns the numerator and denominator

> NUM = [0 2 10 12] and DEN = [1 6 11 6]

and these imply the transfer function

$$H(s) = \frac{2s^2 + 10s + 12}{s^3 + 6s^2 + 11s + 6}$$

which is exactly what we expected. The "1" as the last entry in the statement implies a SISO system and refers to which input – if there are more than one – is of interest.

Example 3.12

Now using the transfer function of Example 3.11, generate a state-space model from it by using the TF2SS statement from MATLAB. We let NUM = [0 2 10 12] and DEN = [1 6 11 6] and write

> [A, B, C, D] = TF2SS(NUM,DEN)

and this returns the matrices

$$A = \begin{bmatrix} -6 & -11 & -6 \\ 1 & 0 & 0 \\ 0 & 1 & 0 \end{bmatrix}$$

$$B = \begin{bmatrix} 1 \\ 0 \\ 0 \end{bmatrix}$$

$$C = \begin{bmatrix} 2 & 10 & 12 \end{bmatrix}$$

$$D_C = \begin{bmatrix} 0 \end{bmatrix}$$

and these are not quite either the OF or the CF as we have defined them. The problem, again, is that we have an infinite number of state-space representations that are valid for a given transfer function.

Another MATLAB expression that might be useful for our purposes is the CANON statement. It will convert a given A, B, C, D representation into either a diagonal form or what is called a "companion" form, which, in fact, is a form with the A matrix in our OF.

Example 3.13

Consider the (A, B, C, D) system that was the output of Example 3.12. Let

        A = [-6 –11 –6; 1 0 0; 0 1 0]
        B = [1 0 0]´
        C = [2 10 12]
        D = [0]

Then [Ab, Bb, Cb, Db] = CANON (A, B, C, D, 'modal') returns the diagonal form

$$A = \begin{bmatrix} -3 & 0 & 0 \\ 0 & -2 & 0 \\ 0 & 0 & -1 \end{bmatrix}$$

$$B = \begin{bmatrix} -4.7697 \\ -4.5620 \\ 0.8660 \end{bmatrix}$$

$$C = \begin{bmatrix} 0 & 0 & 2.3094 \end{bmatrix}$$

$$D = \begin{bmatrix} 0 \end{bmatrix}$$

and [Ab, Bb, Cb, Db] = CANON(A, B, C, D, 'companion') returns the form

$$A = \begin{bmatrix} 0 & 0 & -6 \\ 1 & 0 & -11 \\ 0 & 1 & -6 \end{bmatrix}$$

$$B = \begin{bmatrix} 1 \\ 0 \\ 0 \end{bmatrix}$$

$$C = \begin{bmatrix} 2 & -2 & 2 \end{bmatrix}$$

$$D = \begin{bmatrix} 0 \end{bmatrix}$$

in which we note the A matrix is in the OF.

Next we consider the solution of the state equations. The scalar form of Equation 3.22 is

$$\frac{dx}{dt} = ax + br, \text{ with } x(t_0) = x_0 \tag{3.55}$$

The solution is $x(t) = x_{HO}(t) + x_{FO}(t)$, where $x_{HO}(t) = \overline{K}e^{at}$ and $\overline{K}$ is a constant and $x_{FO}(t)$ can be obtained using the method of variation of parameters. Let $x_{FO}(t) = K(t)e^{at}$, where K is assumed to be time varying. Then we get

$$\frac{dx_{FO}}{dt} = K'e^{at} + aKe^{at} = ax + br = K'e^{at} + ax \tag{3.56}$$

Subtracting the ax term and solving for $\mathbf{K}'$ we get

$$\mathbf{K}' = e^{-at}br \text{ or } K(t) = \int_{-\infty}^{t} e^{-a\lambda} br(\lambda)d\lambda \tag{3.57}$$

Then

$$x_{FO}(t) = \int_{-\infty}^{t} e^{-a(\lambda-t)} br(\lambda)d\lambda \tag{3.58}$$

and

$$x(t) = x_{HO}(t) + x_{FO}(t) = \overline{K}e^{at} + \int_{-\infty}^{t} e^{-a(\lambda-t)} br(\lambda)d\lambda \tag{3.59}$$

Solving for the initial condition, we get

$$x(t_0) = \overline{K}e^{at_0} + \int_{-\infty}^{t_0} e^{-a(\lambda-t_0)} br(\lambda)d\lambda \tag{3.60}$$

and

$$\overline{K} = e^{-at_0} x(t_0) - \int_{-\infty}^{t_0} e^{-a(\lambda)} br(\lambda)d\lambda \tag{3.61}$$

The total solution then becomes

$$x(t) = e^{a(t-t_0)} x(t_0) + \int_{t_0}^{t} e^{at-a\lambda} br(\lambda)d\lambda \tag{3.62}$$

With this solution for the scalar state equation we can generalize and try the following form for the vector/matrix state equation:

$$x(t) = e^{A(t-t_0)} x(t_0) + \int_{t_0}^{t} e^{A(t-\lambda)} Br(\lambda)d\lambda \tag{3.63}$$

Indeed, differentiation and substitution of this equation into Equation 3.22 verifies that it is the solution. Since $x(t_0)$ is assumed to be known and A, B, and r(t) are given, the major problem is to find e^{At} which is sometimes called the *state transition matrix* (STM) $\Phi(t)$:

$$e^{At} = \Phi(t) \tag{3.64}$$

The STM can be written as an infinite series

$$e^{At} = I + At + \frac{1}{2!}(At)^2 + \frac{1}{3!}(At)^3 + \cdots \tag{3.65}$$

that we would like to put into closed form. One way to achieve this is to use the Laplace transform approach. Without zeroing out the initial condition as we did in deriving the transfer function equation, take the Laplace transform of Equation 3.22 assuming for convenience that $t_0 = 0$, solve for $X(s)$, and we get

$$X(s) = (sI - A)^{-1} x(0) + (sI - A)^{-1} BR(s) \tag{3.66}$$

This equation can be written in the time domain as

$$x(t) = L^{-1}[(sI - A)^{-1}]x(0) + L^{-1}[(sI - A)^{-1} BR(s)] \tag{3.67}$$

Comparison of this equation with the solution, Equation 3.63, again with $t_0 = 0$, reveals the following closed-form expression for the STM

$$\Phi(t) = e^{At} = L^{-1}[(sI - A)^{-1}] \tag{3.68}$$

Another way to obtain a closed-form expression for the STM is to employ the Cayley-Hamilton theorem, which says simply that every A matrix satisfies its characteristic equation. A matrix equation is generated out of a scalar equation. We look at the implications of the Cayley-Hamilton theorem in the next example.

Example 3.14

Consider the A matrix

$$A = \begin{bmatrix} -1 & 0 & 0 \\ 0 & -2 & 0 \\ 0 & 0 & -3 \end{bmatrix}$$

and generate its characteristic equation using Equation 3.42:

$$\det[\lambda I - A] = 0 \text{ or } (\lambda + 1)(\lambda + 2)(\lambda + 3) = \lambda^3 + 6\lambda^2 + 11\lambda + 6 = 0$$

This is a scalar equation. The Cayley-Hamilton theorem says that we can replace the scalar λ by the matrix A. That is, $A^3 + 6A^2 + 11A + 6I = 0$, the 3×3 zero matrix, with the constant term 6 multiplied by the identity matrix I. Substituting A into this equation, we can easily verify that it is true. One interesting thing about this equation is that we can solve for A^3 as $A^3 = -6A^2 - 11A - 6I$, which is a function of A^2, A, and the identity matrix. But also, using an iteration process,

$$A^4 = -6A^3 - 11A^2 - 6A = -6(-6A^2 - 11A - 6I) - 11A^2 - 6A = 25A^2 + 60A + 36I$$

is itself a function of A^2, A, and the identity matrix. Similarly, every power of A equals a function of only A^2, A, and the identity matrix. For this example with our 3×3 A matrix, then, we can write the infinite series $e^{At} = \Phi(t)$ as follows:

$$e^{At} = \beta_0(t)I + \beta_1(t)A + \beta_2(t)A^2$$

The results of Example 3.14 can be generalized for any $n \times n$ A matrix and we can write the STM in closed form as follows:

$$e^{At} = \beta_0(t) + \beta_1(t)A + \beta_2(t)A^2 + \cdots + \beta_{n-1}(t)A^{n-1} \qquad (3.69)$$

This is a closed-form expression for the STM, but the time-dependent β terms still need to be determined. Since Equation 3.69 follows from the Cayley-Hamilton theorem, replacing λ by A in the characteristic equation, it is also true that the A equation, Equation 3.69, is satisfied by λ. This can be thought of as a reverse Cayley-Hamilton theorem. It implies another equation, a companion expression,

$$e^{\lambda t} = \beta_0(t) + \beta_1(t)\lambda + \beta_2(t)\lambda^2 + \cdots + \beta_{n-1}(t)\lambda^{n-1} \qquad (3.70)$$

which is a scalar equation in terms of the eigenvalues of the given A matrix. Substituting the eigenvalues into Equation 3.70, we can solve for the β terms, which then are used in Equation 3.69 to determine the STM.

Example 3.15

Using the A matrix

$$A = \begin{bmatrix} -1 & 0 & 0 \\ 0 & -2 & -1 \\ 1 & 0 & -3 \end{bmatrix}$$

determine the STM by the Laplace transform approach and the Cayley-Hamilton approach.

Solution:

$$(sI-A)^{-1} = \begin{bmatrix} s+1 & 0 & 0 \\ 0 & s+2 & 1 \\ -1 & 0 & s+3 \end{bmatrix}^{-1} = \begin{bmatrix} \dfrac{1}{s+1} & 0 & 0 \\ \dfrac{-1}{(s+1)(s+2)(s+3)} & \dfrac{1}{(s+2)} & \dfrac{-1}{(s+2)(s+3)} \\ \dfrac{1}{(s+1)(s+3)} & 0 & \dfrac{1}{(s+3)} \end{bmatrix}$$

and the inverse Laplace transform of this is

$$\Phi(t) = \begin{bmatrix} e^{-t} & 0 & 0 \\ e^{-2t} - 0.5e^{-t} - 0.5e^{-3t} & e^{-2t} & e^{-3t} - e^{-2t} \\ 0.5e^{-t} - 0.5e^{-3t} & 0 & e^{-3t} \end{bmatrix} u(t)$$

For the Cayley-Hamilton method we need the eigenvalues

$$\lambda_1 = -1$$
$$\lambda_2 = -2$$
$$\lambda_3 = -3$$

Then

$$\Phi(t) = e^{At} = \beta_0 I + \beta_1 A + \beta_2 A^2$$

$$
= \begin{bmatrix} \beta_0 & 0 & 0 \\ 0 & \beta_0 & 0 \\ 0 & 0 & \beta_0 \end{bmatrix} + \begin{bmatrix} -\beta_1 & 0 & 0 \\ 0 & -2\beta_1 & -\beta_1 \\ \beta_1 & 0 & -3\beta_1 \end{bmatrix} + \begin{bmatrix} \beta_2 & 0 & 0 \\ -\beta_2 & 4\beta_2 & 5\beta_2 \\ -4\beta_2 & 0 & 9\beta_2 \end{bmatrix}
$$

$$
= \begin{bmatrix} \beta_0 - \beta_1 + \beta_2 & 0 & 0 \\ -\beta_2 & \beta_0 - 2\beta_1 + 4\beta_2 & -\beta_1 + 5\beta_2 \\ \beta_1 - 4\beta_2 & 0 & \beta_0 - 3\beta_1 + 9\beta_2 \end{bmatrix}
$$

Using the scalar equation, we can write

$$
e^{-t} = \beta_0 - \beta_1 + \beta_2
$$

$$
e^{-2t} = \beta_0 - 2\beta_1 + 4\beta_2
$$

$$
e^{-3t} = \beta_0 - 3\beta_1 + 9\beta_2
$$

and to solve for the β terms we can invert a 3×3 matrix using the INV(x) statement from either MATLAB or MATRIX$_x$ and we get

$$
\beta_0 = 3e^{-t} - 3e^{-2t} + e^{-3t}
$$

$$
\beta_1 = 2.5e^{-t} - 4e^{-2t} + 1.5e^{-3t}
$$

$$
\beta_2 = 0.5e^{-t} - e^{-2t} + 0.5e^{-3t}
$$

and finally

$$
e^{At} = \Phi(t) = \begin{bmatrix} \beta_0 - \beta_1 + \beta_2 & 0 & 0 \\ -\beta_2 & \beta_0 - 2\beta_1 + 4\beta_2 & -\beta_1 + 5\beta_2 \\ \beta_1 - 4\beta_2 & 0 & \beta_0 - 3\beta_1 + 9\beta_2 \end{bmatrix}
$$

or

$$\Phi(t) = \begin{bmatrix} e^{-t} & 0 & 0 \\ e^{-2t} - 0.5e^{-t} - 0.5e^{-3t} & e^{-2t} & e^{-3t} - e^{-2t} \\ 0.5e^{-t} - 0.5e^{-3t} & 0 & e^{-3t} \end{bmatrix} u(t)$$

identical to the result from the Laplace transform approach.

Example 3.16

Consider the field-controlled DC motor system of Fig. 3.16. We can find the state-transition matrix $\Phi(t)$ via the Laplace transform approach using Equation 3.68.

$$sI - A = \begin{bmatrix} \dfrac{sL+R}{L} & 0 & 0 \\ 0 & s & -1 \\ -\dfrac{K}{J} & 0 & \dfrac{sJ+f}{J} \end{bmatrix}$$

and

$$[sI - A]^{-1} = \begin{bmatrix} \dfrac{sL+R}{L} & 0 & 0 \\ 0 & s & -1 \\ -\dfrac{K}{J} & 0 & \dfrac{sJ+f}{J} \end{bmatrix}^{-1} = \begin{bmatrix} \dfrac{L}{SL+R} & 0 & 0 \\ \dfrac{KL}{s(sJ+f)(sL+R)} & \dfrac{1}{s} & \dfrac{J}{s(sJ+f)} \\ \dfrac{KL}{(sJ+f)(sL+R)} & 0 & \dfrac{J}{sJ+f} \end{bmatrix}$$

From Appendix C and using partial-fraction expansion, we obtain the inverse Laplace transform of each term in this matrix:

$$\Phi(t) = \begin{bmatrix} e^{\frac{-R}{L}t} & 0 & 0 \\ \dfrac{-\phi}{fR}\{(R-f) - RJe^{\frac{-f}{J}t} + fLe^{\frac{-R}{L}t}\} & 1 & (1 - e^{\frac{-f}{J}t}) \\ \phi\{e^{\frac{-f}{J}t} - e^{\frac{-R}{L}t}\} & 0 & e^{\frac{-f}{J}t} \end{bmatrix} u(t)$$

where $\phi = \dfrac{KL}{RJ - fL}$

So far it should be apparent that all of our state-variable discussion has been with regard to continuous-time systems. But, in fact, our primary emphasis in the book is on discrete systems. Is our stress misplaced? We think not, because knowing the basics of continuous-time systems provides appropriate background for discrete systems since most of the discrete results are available from continuous-time results by simple transformations. Also, most readers of this book will probably have had some grounding in continuous systems and reinforcing that exposure should solidify our point of departure into the discrete domain. Another important consideration is that most of the time the typical physical plant is an energy-based system described by continuous-time mathematics and that given representation then needs to be translated into the digital domain.

State-Variable Description of Discrete Systems

The state-variable and output equations of the standard system in Fig. 3.17 are, respectively

$$x((k+1)T) = Ax(kT) + Br(kT) \tag{3.71}$$

$$c(kT) = Cx(kT) + Dr(kT) \tag{3.72}$$

Figure 3.18   Discrete-time system.

In these equations $x(kT)$ is an n-dimensional state vector, $r(kT)$ is an m-dimensional input vector, $c(kT)$ is a k-dimensional output vector, A is an n x n matrix, B is an n x m matrix, C is a k x n matrix and D is a k x m matrix.

Typically, these equations arise from sampling a system originally given in a continuous-time representation. Repeating the general solution of the continuous-time state equation,

$$x(t) = e^{A(t-t_0)}x(t_0) + \int_{t_0}^{t} e^{A(t-\lambda)}Br(\lambda)d\lambda \tag{3.73}$$

and if we sample and clamp the r(t) at the time intervals kT and let the initial time t_0 be kT, we can write

$$x(t) = e^{A(t-kT)}x(kT) + \int_{Kt}^{t} e^{A(t-\lambda)}Bd\lambda r(kT) \tag{3.74}$$

Letting $t = (k+1)T$, this equation becomes

$$x(kT+T) = e^{AT}x(kT) + [\int_{0}^{T} e^{A\alpha}Bd\alpha]r(kT) \tag{3.75}$$

where the term in brackets follows from a change of variable, namely $\alpha = kT + T - \lambda$, in Equation 3.74. Then if we let

$$\int_{0}^{T} e^{A\alpha}Bd\alpha = \gamma(T) \tag{3.76}$$

and note that the first term to the right of the equals sign in Equation 3.75 is the state-transition matrix evaluated at $t = T$, we can write

$$x((k+1)T) = \phi(T)x(kT) + \gamma(T)r(kT) \tag{3.77}$$

and the continuous-time output equation with the outputs taken at $t = kT$ becomes

$$c(kT) = Cx(kT) + Dr(kT) \tag{3.78}$$

Together these two equations, the first of which is analogous to Equation 3.71 and the second of which is identical to Equation 3.72, constitute the dynamical state-variable representation of a general sampled-data system. Often, the sampling period is normalized to $T = 1$. If the complete system, including plant and controller, is discrete, with normalized sampling period, the representation may take the form

$$x(k+1) = Ax(k) + Br(k) \tag{3.79}$$

and

$$c(k) = Cx(k) + Dr(k) \tag{3.80}$$

In subsequent discussions we will normally use the (A, B, C, D) notation so that our discrete-time results will be analogous to our continuous-time results. Keep in mind that the A and B matrices may need to be replaced by ϕ and γ respectively, depending on whether our systems are sampled-data systems or purely discrete systems.

Example 3.17

A continuous-time system is described by

$$\frac{dx}{dt} = \begin{bmatrix} 0 & 0 \\ 1 & -1 \end{bmatrix} x + \begin{bmatrix} 1 & 1 \\ 1 & -1 \end{bmatrix} r$$

and

$$c = \begin{bmatrix} 1 & 1 \\ 1 & 1 \end{bmatrix} x + \begin{bmatrix} 1 & 1 \\ 1 & 1 \end{bmatrix} r$$

The equivalent discrete-time state model of the system is obtained using Equation 3.76 and the STM equation evaluated at $t = T$: $\phi(t) = L^{-1}[sI - A]^{-1}$ and

$$\begin{bmatrix} s & 0 \\ -1 & s+1 \end{bmatrix}^{-1} = \frac{\begin{bmatrix} s+1 & 0 \\ 1 & s \end{bmatrix}}{s^2 + s}$$

Therefore,

$$\phi(t) = \begin{bmatrix} 1 & 0 \\ 1 - e^{-t} & e^{-t} \end{bmatrix} \text{ and } \phi(T) = \begin{bmatrix} 1 & 0 \\ 1 - e^{-T} & e^{-T} \end{bmatrix} = e^{AT}$$

and

$$\gamma(T) = \int_0^T e^{A\alpha} B d\alpha = \int_0^T \begin{bmatrix} 1 & 1 \\ 1 & 1 - 2e^{-\alpha} \end{bmatrix} d\alpha = \begin{bmatrix} T & T \\ T & T - 2 + 2e^{-T} \end{bmatrix}$$

Matrices $\phi(T)$ and $\gamma(T)$ describe the state equation for the equivalent discrete-time model. If a value for T is stipulated, these matrices become constants and function just like the A and B matrices in the purely discrete equation model.

As in the continuous case, the CF, OF, and JF canonical forms are important. The CF and OF forms are available by reading the coefficients from a discrete system transfer function H(z) represented in the general expression

$$H(z) = \frac{c_{n-1}z^{n-1} + \cdots + c_1 z + c_0}{z^n + a_{n-1}z^{n-1} + \cdots + a_1 z + a_0} + b_n \tag{3.81}$$

The (A, B, C, D) matrices take the same form as in the continuous case. The Jordan form is available with the same procedure as that used in the continuous case. If an arbitrary (A, B, C, D) system description is given, H(z) can be formed from it by taking the z-transform of the state equation. Assuming zero initial conditions and a SISO system, we get

$$H(z) = C(zI - A)^{-1}B + D \tag{3.82}$$

As in the continuous case, to every system there corresponds a unique H(z), but for every system there are infinitely many (A, B, C, D) representations.

Next we present a solution of Equation 3.79 (where, again, it should be understood that A, B will be replaced by ϕ and γ if necessary). Without elaboration we merely note that the difference equation solution for x(k) is analogous to the differential equation solution for x(t). It takes the form

$$x(k) = A^k x(0) + \sum_{n=0}^{k-1} A^{k-n-1} Br(n) \tag{3.83}$$

where A^k is the discrete STM and can be solved for in several ways. As with the continuous STM solution, we will mention only two of these: the z-transform approach and the Cayley-Hamilton method.

The z-transform approach follows from taking the z-transform of Equation 3.79 without first zeroing out the initial condition vector, solving for X(z), then taking the inverse z-transform, and finally comparing terms in this equation to the terms in Equation 3.83. We observe that

$$A^k = z^{-1}[(I - z^{-1}A)]^{-1} \tag{3.84}$$

The Cayley-Hamilton theorem allows us to write

$$A^k = \beta_0(k)I + \beta_1(k)A + \ldots + \beta_{n-1}(k)A^{n-1} \tag{3.85}$$

and

$$\lambda^k = \beta_0(k) + \beta_1(k)\lambda + \ldots + \beta_{n-1}(k)\lambda^{n-1} \tag{3.86}$$

where the λ terms are the eigenvalues of A. As in the continuous case, knowing the eigenvalues, solve Equation 3.86 for the β terms and substitute them into Equation 3.85.

Example 3.18

Let

$$
A = \begin{bmatrix} -1 & 0 & 0 \\ 0 & -2 & -1 \\ 1 & 0 & -3 \end{bmatrix}
$$

and solve for the STM using both the z-transform approach and the Cayley-Hamilton theorem.

Solution:

$$
(I - z^{-1}A)^{-1} = \begin{bmatrix} 1 + z^{-1} & 0 & 0 \\ 0 & 1 + 2z^{-1} & z^{-1} \\ -z^{-1} & 0 & 1 + 3z^{-1} \end{bmatrix}^{-1}
$$

$$
= \frac{\begin{bmatrix} (1 + 2z^{-1})(1 + 3z^{-1}) & -z^{-2} & -z^{-1}(1 + 2z^{-1}) \\ 0 & (1 + z^{-1})(1 + 3z^{-1}) & 0 \\ 0 & -z^{-1}(1 + z^{-1}) & (1 + z^{-1})(1 + 2z^{-1}) \end{bmatrix}^T}{(1 + z^{-1})(1 + 2z^{-1})(1 + 3z^{-1})}
$$

$$
= \begin{bmatrix} \dfrac{1}{1 + z^{-1}} & 0 & 0 \\[3mm] \dfrac{-z^{-2}}{(1 + z^{-1})(1 + 2z^{-1})(1 + 3z^{-1})} & \dfrac{1}{1 + 2z^{-1}} & -\dfrac{z^{-1}}{(1 + 2z^{-1})(1 + 3z^{-1})} \\[3mm] \dfrac{z^{-1}}{(1 + z^{-1})(1 + 3z^{-1})} & 0 & \dfrac{1}{1 + 3z^{-1}} \end{bmatrix}
$$

Then we need to inverse z-transform each term in this expression and finally we get for the STM the following expression:

$$A^k = \begin{bmatrix} (-1)^k & 0 & 0 \\ -0.5(-1)^k + (-2)^k - 0.5(-3)^k & (-2)^k & -(-2)^k + (-3)^k \\ 0.5(-1)^k - 0.5(-3)^k & 0 & (-3)^k \end{bmatrix} u(k)$$

Using the Cayley-Hamilton method we need eigenvalues:

$$\lambda_1 = -1$$

$$\lambda_2 = -2$$

$$\lambda_3 = -3$$

and

$$A^k = \beta_0(k)I + \beta_1(k)A + \beta_2(k)A^2 = \begin{bmatrix} \beta_0 - \beta_1 + \beta_2 & 0 & 0 \\ -\beta_2 & \beta_0 - 2\beta_1 + 4\beta_2 & -\beta_1 + 5\beta_2 \\ \beta_1 - 4\beta_2 & 0 & \beta_0 - 3\beta_1 + 9\beta_2 \end{bmatrix}$$

where the β terms follow from using the eigenvalues in the equation $\lambda^k = \beta_0(k) + \beta_1(k)\lambda + \beta_2(k)\lambda^2$:

$$\beta_0 = 3(-1)^k - 3(-2)^k + (-3)^k$$

$$\beta_1 = 2.5(-1)^k - 4(-2)^k + 1.5(-3)^k$$

$$\beta_2 = 0.5(-1)^k - 2(-2)^k + 0.5(-3)^k$$

Then

$$A^k = \begin{bmatrix} \beta_0 - \beta_1 + \beta_2 & 0 & 0 \\ -\beta_2 & \beta_0 - 2\beta_1 + 4\beta_2 & -\beta_1 + 5\beta_2 \\ \beta_1 - 4\beta_2 & 0 & \beta_0 - 3\beta_1 + 9\beta_2 \end{bmatrix}$$

$$= \begin{bmatrix} (-1)^k & 0 & 0 \\ -0.5(-1)^k + (-2)^k - 0.5(-3)^k & (-2)^k & -(-2)^k + (-3)^k \\ 0.5(-1)^k - 0.5(-3)^k & 0 & (-3)^k \end{bmatrix} u(k)$$

which is the same result obtained with the z-transform method.

Although much more could be said about state-variable representation of linear continuous and discrete systems, we close this discussion with one final observation. Note that the solution of the state equation – in both the discrete and continuous cases – is written as the sum of two terms. This is a general property of linear systems: that the response of the system can be partitioned into two parts. The first term depends on the initial condition $x(0)$ but not on the input. The second term depends on the input but not on the initial condition. The first term is called the *zero-input response* $x_{zi}(t)$ or $x_{zi}(k)$. The second term is called the zero-state response $x_{zs}(t)$ or $x_{zs}(k)$. Then

$$x(t) = x_{zi}(t) + x_{zs}(t) \tag{3.87}$$

and

$$x(k) = x_{zi}(k) + x_{zs}(k) \tag{3.88}$$

This type of decomposition was mentioned in Chapter 2 in the discussion of scalar differential and difference equations. The zero-state response and the zero-input response can be calculated or computed separately. Sometimes only one of these is of interest. Transfer function analysis, for instance, is concerned only with the zero-state response. Also, in the output equation we can associate one part of $c(t)$ or $c(k)$ with the input and one part with the initial conditions. Then

$$c(t) = c_{zi}(t) + c_{zs}(t) \tag{3.89}$$

and

$$c(k) = c_{zi}(k) + c_{zs}(k) \tag{3.90}$$

3.4 NONLINEAR DISCRETE SYSTEMS

All of our discussion so far has been concerned with *linear* continuous and discrete systems. If a system is *nonlinear,* most of the results we have developed will not be very useful. Nonlinear systems theory is an extensive area of discourse in its own right. In this section we present only a glimpse of that discourse, focusing on the technique of linearization and the process of simulation.

As far as linearization is concerned, let us look at a very simple situation, assuming just discrete, time-invariant SISO systems and just the second-order case. Then we can write

$$x(k+1) = f(x(k), r(k)) \qquad (3.91)$$

and

$$c(k) = g(x(k), r(k)) \qquad (3.92)$$

where we assume that the f (a two-dimensional vector) and g (a scalar) nonlinear functions are sufficiently smooth so that we can differentiate them and express them in Taylor series expansions. Then after some manipulation and approximation, the details of which are beyond our scope but are available in several nonlinear texts (e.g., 10), we arrive at the linearized variational equations:

$$x(k+1) = Ax(k) + Br(k) \qquad (3.93)$$

and

$$c(k) = Cx(k) + Dr(k) \qquad (3.94)$$

where the A,B,C,D matrices are called *Jacobian matrices.* They consist of partial derivatives which take the form

$$A = \begin{bmatrix} \dfrac{\partial f_1}{\partial x_1} & \dfrac{\partial f_1}{\partial x_2} \\ \dfrac{\partial f_2}{\partial x_1} & \dfrac{\partial f_2}{\partial x_2} \end{bmatrix} \qquad (3.95)$$

$$B = \begin{bmatrix} \dfrac{\partial f_1}{\partial r} \\ \dfrac{\partial f_2}{\partial r} \end{bmatrix} \tag{3.96}$$

$$C = \begin{bmatrix} \dfrac{\partial g}{\partial x_1} & \dfrac{\partial g}{\partial x_2} \end{bmatrix} \tag{3.97}$$

$$D = [\dfrac{\partial g}{\partial r}]. \tag{3.98}$$

All four of these matrices need to be evaluated at specific operating points which are usually selected to be those points of stationarity at which, for a specified input, $x(k+1) = x(k)$. Then the nature of the system, in the neighborhood of the stationary point or points, is described fairly accurately by the linearized equations. The variables in the linearized equations are really incremental versions of the variables in the given nonlinear equations, but to avoid complicating the notation, we use the same notation in both cases.

Example 3.19

Consider the nonlinear system with zero input:

$$x_1(k+1) = 2x_2(k) - x_1(k)x_2(k)$$
$$x_2(k+1) = x_1(k)$$

If we set the right-hand side of the first equation to $x_1(k)$ and the right-hand side of the second equation to $x_2(k)$, we get the stationary points $(0,0)$ and $(1,1)$. The only Jacobian of interest here is the A matrix:

$$A = \begin{bmatrix} \dfrac{\partial f_1}{\partial x_1} & \dfrac{\partial f_1}{\partial x_2} \\ \dfrac{\partial f_2}{\partial x_1} & \dfrac{\partial f_2}{\partial x_2} \end{bmatrix} = \begin{bmatrix} -x_2(k) & 2 - x_1(k) \\ 1 & 0 \end{bmatrix} = \begin{bmatrix} 0 & 2 \\ 1 & 0 \end{bmatrix}, \begin{bmatrix} -1 & 1 \\ 1 & 0 \end{bmatrix}$$

corresponding to the two stationary points. Both of these A matrices have eigenvalues on the outside of the unit circle in the z-plane, indicating an unstable system.

Now, what about the case where the f and g functions on the right-hand side of Equations 3.91 and 3.92 are not smooth enough to differentiate, for example, if the nonlinearities are things like limiters? Then mathematical analysis becomes tricky if not impossible, and it is usually best to simulate the given system and do a trial-and-error analysis. Fortunately, several CAD tools are now available for simulating complex control systems of the linear and nonlinear as well as continuous, discrete, and sampled-data varieties. For example, MATRX$_X$ uses SYSTEMBUILD and MATLAB uses SIMULINK. We look briefly at the latter. SIMULINK can be called from within MATLAB and allows the user to construct a system in block diagram form by calling on several icons that represent standard blocks, such as transfer functions, summing devices, gain multipliers, and saturation devices. All one need do is connect the blocks together, apply the proper input, and run the simulation. After looking at the outputs it is very easy to change things in the system, such as a gain value, and do another simulation. Some examples will illustrate the power of SIMULINK.

Example 3.20

Consider a continuous-time unity gain feedback system with open-loop transfer function $G(s) = \dfrac{1}{(s^2 + s + 1)}$ driven by a unit step input. We know from simple analysis that the closed-loop transfer function is

$$\frac{G}{1+G} = \frac{1}{s^2 + s + 2}$$

which has an undamped natural frequency of 1.414 and a damping ratio of 0.353 indicating an under-damped response. The system is set up in SIMULINK as indicated in Fig. 3.19. After applying the unit step input (which has a 1-second delay) we get the output plot that we might expect, as shown in Fig. 3.20.

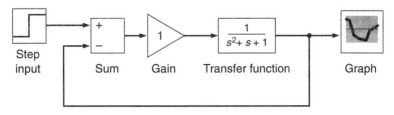

Figure 3.19 SIMULINK block diagram for Example 3.20.

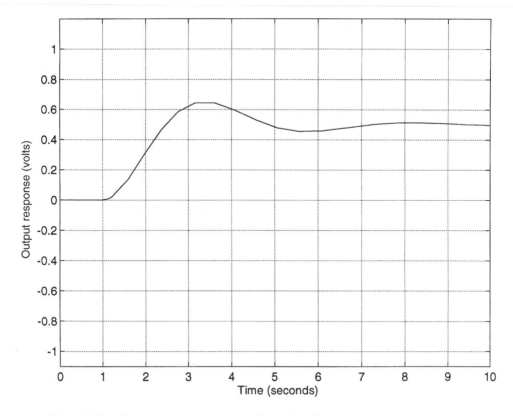

Figure 3.20   Output response of system in Example 3.20.

Example 3.20 was quite simple to analyze and we did not really need the simulation. But when nonlinearities are involved, the analysis can quickly get out of hand. In the next example we look at a sampled-data system that is linear to start with, but then we add a saturation or limiting device and examine the results.

Example 3.21

Consider a unity gain feedback sampled-data system with plant $G(s) = \frac{1}{(s+1)}$ and discrete controller

$$G(z) = \frac{2(z+0.25)(z+0.75)}{z(z-0.5)}$$

driven by a unit step input. The SIMULINK block diagram appears as in Fig. 3.20.

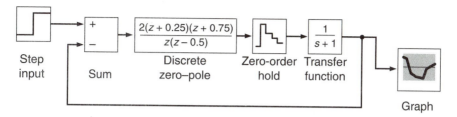

Figure 3.20   SIMULINK block diagram for Example 3.21.

The zero-order-hold device is actually not needed in the simulation because every discrete transfer function in SIMULINK is preceded automatically by a sampler and followed by a ZOH device. We do have to enter a value for T, the sampling period. A value of T = 0.2 second produces the output shown in Fig. 3.21, indicating quite a lot of oscillation. If we increase T to 0.3 second the oscillation gets worse, so we decrease it to T = 0.1 and the output looks fairly good. These results are shown in Fig. 3.22 and 3.23, respectively.

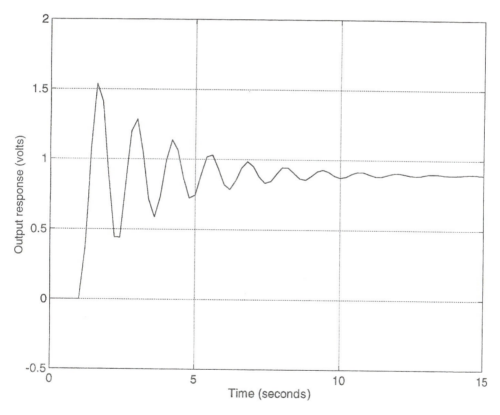

Figure 3.21   Output of Example 3.21 with T = 0.2 second.

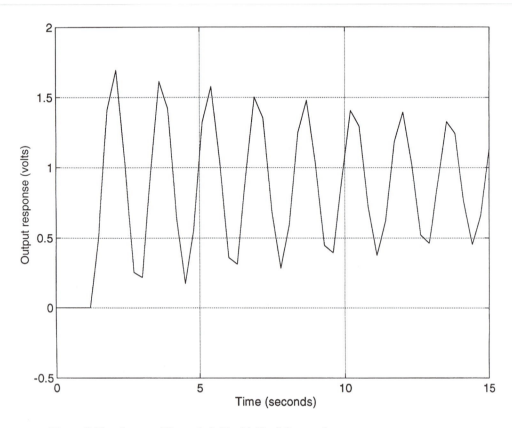

Figure 3.22   Output of Example 3.21 with T = 0.3 second.

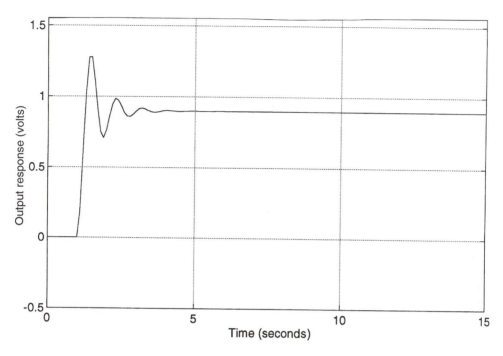

Figure 3.23   Output of Example 3.21 with T = 0.1 second.

Now if in the same system we introduce a saturation device in the place where the ZOH had been, the SIMULINK block diagram appears as in Fig. 3.24. We adjust the saturation device so that it limits at +1 on the top and -1 on the bottom. Letting T = 0.3, we get a little ripple in the response which dies out eventually, as indicated in Fig. 3.25. If T = 0.5 the response is as in Fig. 3.27, indicating the presence of a small limit cycle. A general conclusion here is that if the sampling period needs to be large, using the nonlinear element can keep the responses from getting too wild. However, driving the system into its limiting modes may be inadvisable because it may cause too much wear and tear on system components. As usual, engineering judgment should have the final word in these matters.

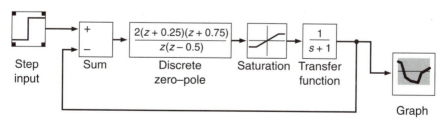

Figure 3.24   SIMULINK block diagram of nonlinear sampled-data system.

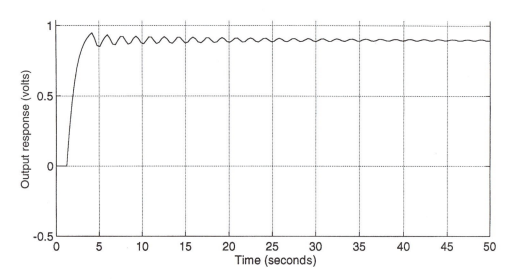

Figure 3.25   Nonlinear system response with T = 0.3 second.

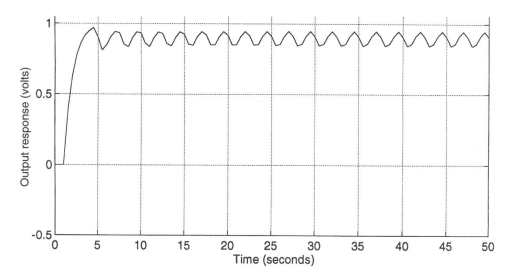

Figure 3.26   Nonlinear system response with T = 0.5 second.

To wrap up this brief nonlinear discussion, we present an example of a simple first-order system that yields *chaos* under some circumstances. Chaos is a popular idea that has pervaded many dimensions of culture.   The metaphor is potent.   One reason for the explosive growth of the idea stems from nonlinear systems theory. For the last few decades much research has produced many interesting results in the field. No doubt everyone has seen the beautiful computer-generated images of fractal and chaotic systems. The basic idea about chaos is that some deterministic systems, if they are nonlinear, can yield outputs that look like noise but in fact have a very complex structure that can be analyzed from several points of view, and this structure has come to be called chaos. This noiselike appearance is sometimes called *quasi-stochastic*.

In a purely linear system, the essential dynamic nature of the system can be described by the system function. The poles and zeros tell the tale. No reference need be made to the input. But in a nonlinear system, the dynamics can exhibit considerable variety depending on the input. Changing initial conditions will not affect the essential behavior of a linear system, but can cause drastic shifts in nonlinear system behavior. Also, slight changes in a system parameter will generally cause slight changes in a linear system behavior but can result in drastic changes in a nonlinear system behavior. For continuous-time systems at least a second-order system is needed to exhibit chaotic behavior, but for discrete-time systems a first-order nonlinear system will suffice. The next and final example in this section illustrates how a certain first-order discrete-time nonlinear system, under the right circumstances, becomes chaotic.

Example 3.22

Consider the following first-order system:

$$y(k) = ay(k-1)[1 - y(k-1)]$$

and consider varying the parameter a from 2.5 to 3.9, keeping the initial condition fixed at 0.25. The SIMULINK block diagram is presented in Fig. 3.27 (with a = 3.9) and the plots of y(k) for a = 2.5, 3.2, 3.5, and 3.9 are presented in Fig. 3.28, 3.29, 3.30, and 3.31, respectively.

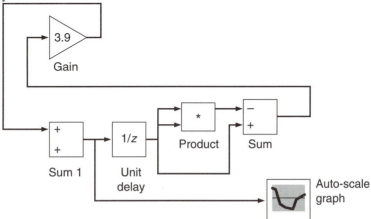

Figure 3.27 SIMULINK block diagram for Example 3.22.

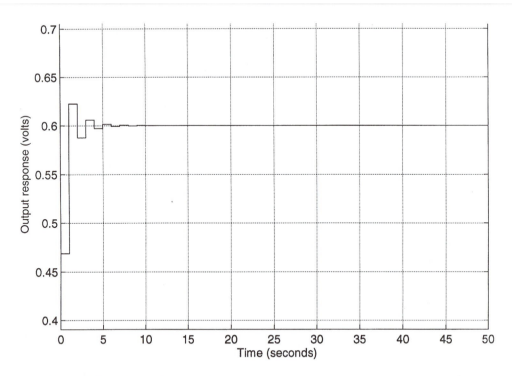

Figure 3.28   Plot of y(k) for a = 2.5.

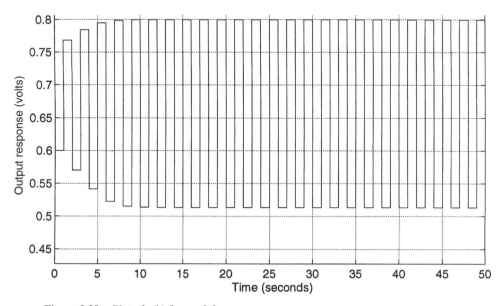

Figure 3.29   Plot of y(k) for a = 3.2.

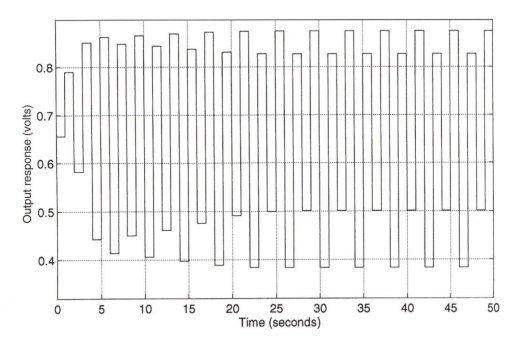

Figure 3.30   Plot of y(k) for a = 3.5.

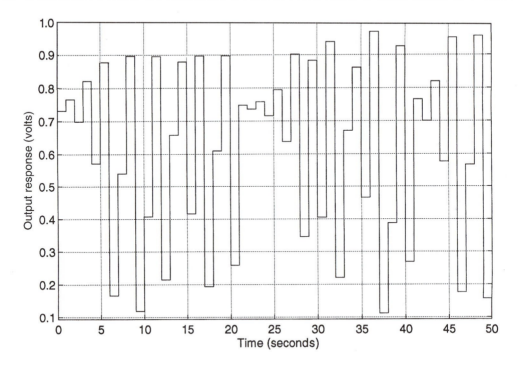

Figure 3.31   Plot of y(k) for a = 3.9.

We can note that for a = 2.5 the response is underdamped but settles out quickly to a value around 0.6. For an a value of 3.2 the response is oscillatory and, for a = 3.5 it is also oscillatory but of a slightly more complex nature. A small increase of the a value to 3.9 produces a bounded nonperiodic response that is, in fact, chaotic. At that we leave this brief discussion of nonlinear systems and turn to another brief discussion of a topic that has also filled volumes, namely, stability.

3.5 Stability Analysis

In this section we study the stability of linear time-invariant discrete systems. Consider such a system shown in Fig. 3.32 with the output being the zero-state response

$$c_{zs}(k) = \sum_{m=0}^{k} h(k-m)r(k) \tag{3.99}$$

For every bounded input sequence r(k), the output sequence $c_{zs}(k)$ is bounded if and only if

$$\sum_{k=0}^{\infty} |h(k)| < \infty \qquad (3.100)$$

and if this expression is to be satisfied, then $h(k)$ must consist of terms of the form a^k where $|a| < 1$. The z-transform of an $h(k)$ expression consisting of such a^k terms will consist of $z/(z-a)$ factors, which means that a linear time-invariant discrete system is Bounded-Input-Bounded-Output (BIBO) stable if and only if:

1. $H(z)$ is a rational function of z.
2. Each pole of $H(z)$ has a magnitude less than unity. [This means that all of the poles of $H(z)$ are inside the unit circle.]

Any nonrepeated poles on the unit circle indicate a response term that stays final but does not die out as $k \to \infty$. Such a system is said to be *marginally stable*.

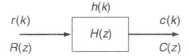

Figure 3.32   Linear time-invariant discrete-time system.

We can also study stability of linear time-invariant discrete systems in terms of the dynamical equations (Equations 3.79 and 3.80) that describe such systems. For the zero-input response ($r = 0$), the state equation reduces to

$$x(k+1) = Ax(k) \qquad (3.101)$$

and the system stability is determined from the eigenvalues of the A matrix in the state equation. In fact, the zero-input response of the system described by Equation 3.101 is *asymptotically stable* if and only if all the eigenvalues of A have magnitudes less than unity. But since the eigenvalues of A are the poles of $H(z)$, asymptotic stability and BIBO stability amount to the same thing.

With slight changes, we can use many continuous-domain stability analysis techniques in the discrete domain. These include Bode plots and the Routh-Hurwitz method. The necessary changes arise from the fact that the stability region for a discrete system differs from its continuous counterpart. By using the bilinear transformation (introduced in Chapter 2) we can transform from the z-plane into the w-plane and therefore change the stability region from the unit circle of the z-plane into the imaginary axis of the w-plane.

$$z = \frac{1 + \dfrac{T}{2}w}{1 - \dfrac{T}{2}w} \qquad (3.102)$$

or

$$w = \frac{2}{T}\frac{z-1}{z+1} \qquad (3.103)$$

This transformation enables us to use the familiar techniques of continuous systems analysis, such as the Routh-Hurwitz method, in the discrete domain.

Routh-Hurwitz Stability Method

The Routh-Hurwitz stability method provides a quick check for stability of continuous systems. This method determines whether any roots of a continuous system's characteristic equation lie in the right half of the s-plane without knowing the exact locations of the roots. To use the Routh-Hurwitz method, complete the following steps:

1.   Find the system's characteristic equation. The characteristic polynomial for a continuous-time closed-loop system has the form

$$q(s) = p_0 + p_1 s + p_2 s^2 + \cdots + p_n s^n \qquad (3.104)$$

2.   Form the Routh-Hurwitz array

$$
\begin{array}{c|cccc}
s^n & p_n & p_{n-2} & p_{n-4} & \cdots \\
s^{n-1} & p_{n-1} & p_{n-3} & p_{n-5} & \cdots \\
s^{n-2} & a_1 & a_2 & a_3 & \\
s^{n-3} & b_1 & b_2 & b_3 & \\
\vdots & \vdots & & & \\
s^0 & n_1 & & &
\end{array}
$$

where the a_i coefficients are from the determinant of cross multiplications of the first two rows. For example:

$$a_1 = \frac{p_{n-1}p_{n-2} - p_n p_{n-3}}{p_{n-1}}, \; a_2 = \frac{p_{n-1}p_{n-4} - p_n p_{n-5}}{p_{n-1}} \qquad (3.105)$$

and

$$b_1 = \frac{a_1 p_{n-3} - a_2 p_{n-1}}{a_1}, \; b_2 = \frac{a_1 p_{n-5} - a_3 p_{n-1}}{a_1} \qquad (3.106)$$

3. Count the number of sign changes for the terms in the first column. This is equal to the number of roots with positive real parts.

Example 3.23

Consider the characteristic equation

$$q(s) = s^3 + 4s^2 + 10s + 10$$

Forming the Routh-Hurwitz array

$$
\begin{array}{c|cc}
s^3 & 1 & 10 \\
s^2 & 4 & 10 \\
s^1 & 7.5 & 0 \\
s^0 & 10 &
\end{array}
$$

We note that there are no changes of sign in the first column: therefore, the RHP has no roots and the system is stable.

Example 3.24

Consider the characteristic equation

$$q(s) = 2s^4 + 2s^3 + s^2 + 10s + 10$$

Forming the Routh-Hurwitz array

$$
\begin{array}{c|ccc}
s^4 & 2 & 1 & 10 \\
s^3 & 2 & 10 & 0 \\
s^2 & a_1 & a_2 & a_3 \\
s^1 & b_1 & b_2 & b_3 \\
s^0 & c_1 & c_2 & c_3
\end{array}
$$

we calculate $a_1 = -9$, and because there is a sign change in the first column, the system is unstable.

There are two special cases that occasionally occur with the Routh-Hurwitz method: A zero may appear in the first column or a row of zeros may appear. If a zero appears in the first column, division by it will of course be problematic. The typical way (but not the only way) to deal with this situation is to replace the zero by $\varepsilon > 0$ and continue the array. Then take the limit as $\varepsilon \to 0$ and we should be able to make the proper determinations.

Example 3.25

Consider the characteristic equation

$$
q(s) = s^4 + 2s^3 + 3s^2 + 6s + 6
$$

Forming the Routh-Hurwitz array

$$
\begin{array}{c|ccc}
s^4 & 1 & 3 & 6 \\
s^3 & 2 & 6 & 0 \\
s^2 & a_1 & a_2 & a_3 \\
s^1 & b_1 & b_2 & b_3 \\
s^0 & c_1 & c_2 & c_3
\end{array}
$$

we calculate $a_1 = 0$ and $a_2 = 6$ and $a_3 = 0$. To deal with the division by zero that would be required in the b-row, we let $a_1 = \varepsilon$. Then $b_1 = \dfrac{(6\varepsilon - 12)}{\varepsilon}$, $b_2 = 0$, and $b_3 = 0$. Also, $c_1 = 6$. The table becomes

$$
q(w) = q\left(z = \frac{1+w}{1-w}\right) = 0
$$

and in the limit as ε gets very small, the b_1 term becomes negative, indicating two sign changes. The system is therefore unstable.

If an entire row of zeros appears, refer to the previous row and form an auxiliary equation from it. An auxiliary equation is a polynomial equation [A(s) = 0] whose order is the order of the row being referenced. The coefficients in the polynomial are the coefficients in the row. For instance, if the last example had yielded all zeros in the a-row, the auxiliary equation would be $A(s) = 2s^3 + 6s = 0$. The auxiliary equation is a factor of the original polynomial and its solution gives us some of the $j\omega$-axis roots of the original polynomial. To continue the Routh-Hurwitz array, differentiate the auxiliary equation and use the resulting coefficients in place of the row of zeros. The next example will illustrate this procedure.

Example 3.26

Consider the polynomial

$$q(s) = s^6 + 3s^5 + 5s^4 + 9s^3 + 8s^2 + 6s + 4$$

Forming the Routh-Hurwitz array

$$
\begin{array}{c|cccc}
s^6 & 1 & 5 & 8 & 4 \\
s^5 & 3 & 9 & 6 & 0 \\
s^4 & 2 & 6 & 4 & 0 \\
s^3 & 0 & 0 & 0 & 0 \\
s^2 & & & & \\
s^1 & & & & \\
s^0 & & & &
\end{array}
$$

and the row of zeros in the third row implies the auxiliary equation from the preceding row: $A(s) = 2s^4 + 6s^2 + 4 = 0$. Differentiating, we get the row 8 12 0, which can be substituted for the row of zeros and we continue the array:

$$
\begin{array}{c|cccc}
s^6 & 1 & 5 & 8 & 4 \\
s^5 & 3 & 9 & 6 & 0 \\
s^4 & 2 & 6 & 4 & 0 \\
s^3 & 8 & 12 & 0 & 0 \\
s^2 & 3 & 4 & 0 & 0 \\
s^1 & 1.33 & 0 & 0 & 0 \\
s^0 & 4 & 0 & 0 & 0
\end{array}
$$

noting that there are no changes of sign in the first column, therefore there are no poles in the RHP. However, the auxiliary equation does provide us with roots of the original polynomial. A(s) is factorable (after dividing by 2) into the form $(s^2 + 2)(s^1 + 1) = 0$, and from this we get poles at $\pm j$ and $\pm j\sqrt{2}$, which indicates a marginally stable system.

Now, given the ease with which we can factor poynomials, say in MATLAB, a question arises as to why use the Routh-Hurwitz criterion at all. The prime reason is that often we are interested in adjusting a parameter in a system and we would like to know the range over which such adjustment yields a stable system. The next example illustrates this.

Example 3.27

Consider the characteristic polynomial

$$
q(s) = s^4 + 2s^3 + 3s^2 + 5s + K
$$

The Routh-Hurwitz array becomes

$$
\begin{array}{c|ccc}
s^4 & 1 & 3 & K \\
s^3 & 2 & 5 & 0 \\
s^2 & 0.5 & K & 0 \\
s^1 & \dfrac{2.5 - 2K}{0.5} & 0 & 0 \\
s^0 & K & 0 & 0
\end{array}
$$

and in order that all the entries in the first column be positive, it is necessary that $K > 0$ and $2.5 - 2K > 0$ or that $0 < K < 1.25$. This is the range of parameter variation over which the system will be stable.

The Routh-Hurwitz method cannot be used directly in the discrete domain because it is based on information from the right half of the s-plane. But by using the bilinear transformation, Equations 3.102 and 3.103, we can apply the Routh-Hurwitz criterion with a new w-variable. The procedure used to form the Routh-Hurwitz array can be used in the discrete domain with the characteristic equation formed with the w-variable. Since Equations 3.102 and 3.103 are valid for any T value, but for illustrative purposes in the next example we let T = 2 and use the mapping

$$z = \frac{1+w}{1-w} \tag{3.107}$$

This mapping transforms the unit-circle of the z-plane into the right half of the s-plane, allowing the Routh-Hurwitz criterion to be used for discrete characteristic equations.

Example 3.28

Let

$$q(z) = z^4 - 1.5z^3 + z^2 - 0.5z + 0.5 = 0$$

be the characteristic equation of a discrete system. To use the Routh-Hurwitz method, form

$$q(w) = q\left(z = \frac{1+w}{1-w}\right) = 0$$

After much tedious algebra we get a complex expression whose numerator set equal to zero is

$$4.5w^4 + 4w^3 + 7w^2 + 0.5 = 0$$

and then forming the Routh-Hurwitz array

w^4	4.5	7	0.5
w^3	4	0	0
w^2	7	0.5	0
w^1	$-\frac{2}{7}$	0	0
w^0	0.5	0	0

we note that there are two changes of sign in the first column, so the system is unstable.

This procedure generally requires much tedious algebra and if only characteristic equation roots are needed, we prefer to use just the ROOTS(X) statement in either

MATLAB or MATRIX$_X$. In Example 3.28, for instance, looking at the original z-domain polynomial we can write X=[1 -1.5 1 -0.5 0.5] as the characteristic equation coefficients and ROOTS(X) from MATRIX$_X$ yields the four roots $z = 0.15160 \pm 0.6611j$ and $q(1)\rangle 0$. The last two roots are outside the unit circle in the z-plane implying an unstable system. However, as we mentioned earlier, often the most important use of the Routh-Hurwitz procedure is to determine the stability range of a system parameter. In that case the procedure above might still be useful. But there is another stability test we can employ for discrete systems, a procedure called the *Jury test*.

Jury Test

The Jury stability method determines the necessary and sufficient conditions for the roots of the characteristic equation of a system to lie inside the unit circle. It is a tabular method similar to the Routh-Hurwitz method used in the continuous domain. The characteristic equation, q(z), of a discrete system is

$$q(z) = \alpha_n z^n + \alpha_{n-1} z^{n-1} + \ldots + \alpha_1 z + \alpha_0 = 0 \tag{3.108}$$

where $\alpha_0, \alpha_1, \cdots, \alpha_n$ are real coefficients and we assume that the first coefficient is positive or can be made so by multiplication of the polynomial by -1.

We use the following procedure to determine the stability of a system using the Jury test.

1. Check conditions a through c:

    a. $q(1)\rangle 0$ (3.109)

    b. $q(-1)\rangle 0$ for n even   (3.110)

    or

    $q(-1)\langle 0$ for n odd   (3.111)

    c. $|\alpha_0|\langle \alpha_n$ (3.112)

    If all the conditions above are satisfied, go to step 2. Table 3.2 shows how to construct the Jury array based on the information from the characteristic equation, q(z), Equation 3.108.

2. Make the Jury array using Table 3.2,

Table 3.2 General Form of the Jury Array

z^0	z^1	z^2	$\cdots$	z^{n-2}	z^{n-1}	z^n
α_0	α_1	α_2	$\cdots$		α_{n-1}	α_n
α_n	α_{n-1}	α_{n-2}	$\cdots$		α_1	α_0
β_0	β_1	β_2	$\cdots$		β_{n-1}	
β_{n-1}	β_{n-2}	β_{n-3}	$\cdots$		β_0	
γ_0	γ_1	γ_2	$\cdots$	γ_{n-2}		
γ_{n-2}	γ_{n-3}	γ_{n-4}	$\cdots$	γ_0		
$\vdots$	$\vdots$	$\vdots$				
η_0	η_1	η_2	η_3			
η_3	η_2	η_1	η_0			
υ_0	υ_1	υ_2				

where

$$\beta_n = \begin{vmatrix} \alpha_0 & \alpha_{n-k} \\ \alpha_n & \alpha_k \end{vmatrix} \quad \text{for k = 0,1,2,...,n-1} \tag{3.113}$$

and

$$\gamma_k = \begin{vmatrix} \beta_0 & \beta_{n-1-k} \\ \beta_{n-1} & \beta_k \end{vmatrix} \quad \text{for k = 0,1,2,...,n-2.} \tag{3.114}$$

Check the following conditions:

$$\begin{aligned} |\beta_0| &> |\beta_{n-1}| \\ |\gamma_0| &> |\gamma_{n-2}| \\ &\vdots \qquad \vdots \\ |\upsilon_0| &> |\upsilon_2| \end{aligned} \tag{3.115}$$

The system is stable if and only if the conditions in Steps 1 and 2 are satisfied.

Example 3.29

The characteristic equation of a discrete-time system is given by

$$F(z) = z^2 + z + 10$$

Applying the Jury test, we have

    a. $F(1) = 12 > 0$

    b.   For n = 2 (even), $F(-1) = 10 > 0$

    c. $|\alpha_0| = 10$ and $|\alpha_n| = 1$ for the characteristic equation. Since the condition $|\alpha_0| < \alpha_0$ is not satisfied, the system is not stable.

Example 3.30

Determine the range of parameter a such that the system represented by the following discrete characteristic equation is stable:

$$F(z) = z^3 + az^2 + az - 0.1$$

Applying the Jury test, we have

    a. $\begin{array}{l} F(1)\rangle 0 \\ a\rangle - 0.45 \end{array}$

    b. $\begin{array}{l} F(-1)\langle 0 \\ a\rangle - 0.45 \end{array}$

    c. $= \dfrac{\partial T}{\partial P}\dfrac{P}{T}$

Since the conditions above are satisfied, we can construct the Jury array:

z^0	z^1	z^2	z^3
-0.1	a	a	1
1	a	a	-0.1
-0.99	-1.1a	-1.1a	

Then the condition $\left|\upsilon_0\right| > \left|\upsilon_2\right|$ implies that $0.99 > 1.1a$ or $a < 0.9$. Therefore, $-0.45 < a < 0.9$ gives us our range of stability.

As with the Routh-Hurwitz procedure, the Jury test occasionally encounters the situation where a zero or row of zeros occurs in the table. The techniques of dealing with these special cases in the Jury test are similar to but a bit more involved than those in the Routh-Hurwitz procedure and are omitted here. Instead, we will move on to another stability technique, due to Liapunov, that can be employed in a wider class of systems than the Routh-Hurwitz and Jury procedures.

Liapunov Stability Method

Assuming that our systems are linear and if we have the system closed-loop transfer function available, the previous methods are effective in stability determinations. All we need to do is consider the denominator of the closed-loop transfer function set equal to zero to get the characteristic equation. If we have the state-variable description (A,B,C,D), of course, we can generate the characteristic equation via

$$\left|\lambda I - A\right| = 0 \tag{3.116}$$

However, it is possible to work directly with the system A matrix and use the Liapounv technique to investigate asymptotic system stability. Unlike the stability methods discussed previously, the Liapunov technique can be used for continuous and discrete as well as linear, nonlinear, and time-varying systems. It is based on the idea that if the energy of a system decreases with time, the system is stable.

Let us look first at linear discrete systems. The Liapunov stability method employs the notion of *positive definite* (PD) functions or matrices. As we indicate in Appendix E, positive definiteness for a matrix simply means that all of its eigenvalues are positive. A function is PD if and only if it is greater than zero for all values of its argument, except possibly at zero, where the function may be zero. $V(x) = x^2$ is an example of a PD function. The major difficulty with this stability approach is finding a Liapunov function that suits the system.

Definition 3.2 For a linear system, a Liapunov function (which is a function of the state vector x) is defined as

$$V = x^T M x \tag{3.117}$$

where M is a positive definite matrix.

Theorem 3.1 A linear discrete-time system with A matrix from its dynamical Equation (3.79) is asymptotically stable if and only if

$$M - A^TMA = N \qquad (3.118)$$

is satisfied for any PD, symmetric, real matrix N. It is usually simplest to choose N as the identity matrix. Then, given A, solve for M and check it for positive definiteness. If it is PD, the system is stable; if not PD, the system is unstable.

Example 3.31

A homogeneous discrete-time system has the dynamical equation

$$x(k+1) = \begin{bmatrix} 1 & 5 \\ -5 & 10 \end{bmatrix} x(k)$$

Let

$$N = \begin{bmatrix} 1 & 0 \\ 0 & 1 \end{bmatrix} \text{ and } M = \begin{bmatrix} a & b \\ c & d \end{bmatrix}$$

Substituting into Equation 3.118 yields

$$\begin{bmatrix} a & b \\ c & d \end{bmatrix} - \begin{bmatrix} 1 & -5 \\ 5 & 10 \end{bmatrix} \begin{bmatrix} a & b \\ c & d \end{bmatrix} \begin{bmatrix} 1 & 5 \\ -5 & 10 \end{bmatrix} = \begin{bmatrix} 1 & 0 \\ 0 & 1 \end{bmatrix}$$

$$\begin{bmatrix} 5b + 5c - 25d & -5a - 9b + 25c + 50d \\ -5a + 25b - 9c + 50d & -25a - 50b - 50c - 99d \end{bmatrix} = \begin{bmatrix} 1 & 0 \\ 0 & 1 \end{bmatrix}$$

$$M = \begin{bmatrix} -0.10803 & 0.04055 \\ 0.04055 & -0.02377 \end{bmatrix}$$

which is not a positive definite matrix. Therefore the system is not asymptotically stable.

Both MATLAB and MATRIX$_X$ have statements that will solve the discrete Liapunov Equation 3.118. The MATLAB statement is M=DLYAP(A',N) which will solve the Liapunov equation for a given A matrix (note the transpose in the MATLAB statement), and specified N matrix (usually chosen as the identity).

Example 3.32

Let

$$A = \begin{bmatrix} 1 & 3 \\ -3 & -2 \end{bmatrix}$$

and determine whether the system x(k+1)=Ax(k) is asymptotically stable.

Solution: Let A=[1 3; -3 -2] and N=[1 0;0 1] or eye(2), which is the MATLAB expression for the identity matrix. Then M=DLYAP(A',N) returns

$$M = \begin{bmatrix} -0.2857 & -0.1905 \\ -0.1905 & -0.2381 \end{bmatrix}$$

which is not positive definite and the system is unstable.

Example 3.33

Now let

$$A = \begin{bmatrix} -0.5 & 0 & 0 \\ 0 & -0.25 & 0 \\ 0 & 0 & 0.25 \end{bmatrix}$$

which is an obviously stable system since the eigenvalues are all inside the unit circle. Checking with MATLAB, letting N=eye(3), we get

$$M = \begin{bmatrix} 1.3333 & 0 & 0 \\ 0 & 1.0667 & 0 \\ 0 & 0 & 1.0667 \end{bmatrix}$$

which is positive definite and the system is stable.

For nonlinear systems the Liapunov method is more difficult, but it can be useful if a proper Liapunov function is chosen.

Theorem 3.2 A nonlinear, autonomous, discrete-time system $x(k+1) = f(x(k))$ is asymptotically stable at the origin if and only if there exists a positive definite (PD) Liapunov function $V(x(k))$ for which $\Delta V(x(k)) = V(x(k+1)) - V(x(k))$ is negative definite (ND), which means that its value is never positive for any value of its argument but may be zero if its argument is zero.

Example 3.34

Consider the following nonlinear system:

$$x_1(k+1) = 0.5x_2(k)$$
$$x_2(k+1) = 0.2x_1(k)x_2(k)$$

If we choose the Liapunov function to be $V(x) = \alpha x_1(k)^2 + \beta x_2(k)^2$ where the constants α and β are real and positive, the Liapunov function is PD but we must look at the variational function $\Delta V(x(k)) = V(x(k+1)) - V(x(k))$ from which we construct

$$\Delta V(x(k)) = V(x(k+1)) - V(x(k))$$
$$= \alpha x_1(k+1)^2 + \beta x_2(k+1)^2 - \alpha x_1(k)^2 - \beta x_2(k)^2$$
$$= \alpha 0.25x_2(k)^2 + \beta 0.04x_1(k)^2 x_2(k)^2 - \alpha x_1(k)^2 - \beta x_2(k)^2$$

and we want this to be ND.   Setting $\beta = 1$ and $\alpha = 4$, we get $\Delta V(x(k)) = 0.04x_1(k)^2 x_2(k)^2 - 4x_1(k)^2 = x_1(k)^2(0.04x_2(k)^2 - 4)$, which is ND as long as $|x_2(k)| < 10$.   The strip in the $x_2(k), x_1(k)$ state space, $-10 < x_2(k) < 10$, represents a stable region.   That is, as long as $x_2(k)$ remains bounded for all values of k, then all initial conditions in the stable region indicated will eventually die out to zero.   However, from the state equations, it appears that we need $|x_2(k)| < 10$ *and* $|x_1(k)| < 5$, or else the state values will grow as time progresses.   In fact, what we actually need is that the initial conditions lie inside an ellipse $(4x_1^2 + x_2^2 < 100)$ to guarantee asymptotic stability in the sense of Liapunov.

The Liapunov method for nonlinear systems is very difficult to apply because of the need to show that the variational expression is ND.   The example we considered was difficult enough.   For a higher-order system the method becomes untenable.   In these cases the best approach is probably simulation, looking at several of the kinds of inputs and initial conditions one would be expected to see.

3.6 SENSITIVITY ANALYSIS

When a transfer function parameter deviates from its nominal value, we need to measure the effect of this change on the overall system transfer function. Ideally we would like our systems to be *robust* or impervious to parameter deviations and external disturbances. For example, in a thermal heating system, the system transfer function is a function of thermal resistance of insulation (R). In such a system we need to measure the sensitivity of the system transfer function as R changes due to environmental effects. The sensitivity of a transfer function T, which may be an open- or closed-loop transfer function, with respect to the parameter P is

$$S_P^T = \frac{\partial T/T}{\partial P/P} = \frac{\partial T}{\partial P}\frac{P}{T} \qquad (3.119)$$

Example 3.35

Fig. 3.33 shows a closed-loop discrete system where the overall closed-loop transfer function is

$$M(z) = \frac{C(z)}{R(z)} = \frac{G(z)}{1+G(z)}$$

$$M(z) = \frac{\dfrac{k}{z^2+0.5z}}{1+\dfrac{k}{z^2+0.5z}} = \frac{k}{z^2+0.5z+k}$$

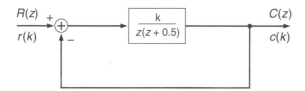

Figure 3.33 Closed-loop discrete system.

From Equation 3.119 we measure the sensitivity of the transfer function due to changes in k.

$$S_K^M = \frac{\partial M}{\partial K}\left\{\frac{K}{M}\right\}$$

$$S_K^M = \frac{z^2 + (0.5 - 2k)z + 0.5k}{z^2 + 0.5z + k}$$

In this system we are interested in how a small change in gain k affects the overall closed-loop transfer function. But let us assume that the frequency of operation is one-fourth of the sampling frequency, that is, $\omega = \omega_s / 4$. Then $z = e^{j\omega T}$ for the frequency response and $z = e^{j\omega T} = e^{jT\omega_s / 4} = e^{j\pi/2} = j$, so the sensitivity function becomes

$$S_K{}^M = \frac{z^2 + (0.5 - 2k)z + 0.5k}{z^2 + 0.5z + k} = \frac{-1 + (0.5 - 2k)j + 0.5k}{-1 + 0.5j + k}$$

This function has a magnitude $\dfrac{\sqrt{(0.5k - 1)^2 + (0.5 - 2k)^2}}{\sqrt{(1 - k)^2 + 0.25}}$ which can be plotted as a function of k using the MATLAB statement PLOT. We enter the magnitude function, calling it y, and then let $x = k$ take the range from -0.25 to 1.00 in steps of 0.05 with the statementk = [-0.25:0.05:1.00].   Then PLOT(k,y) will yield the plot in Fig. 3.34. But we know from stability considerations that the given system will be stable for k values in the range -0.25 < k < 1.0. If we decide, for example, that the k value should be in the range $0 < k < 0.5$, the best value for minimal sensitivity is approximately k = 0.2.

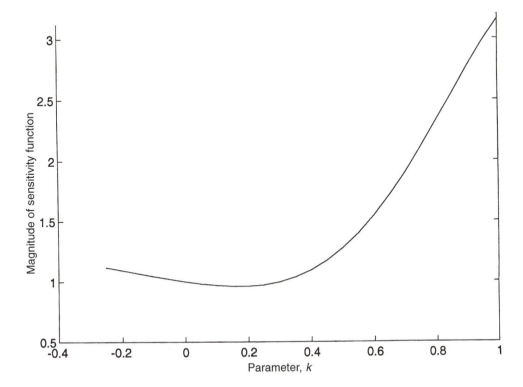

Figure 3.34   MATLAB plot for Example 3.35.

Example 3.36

Fig. 3.35 shows another closed-loop discrete system in which we are interested in the sensitivity of the system to a variation in a parameter:

$$Sa^M = \frac{\partial M}{\partial a} \{\frac{a}{M}\}$$

$$M(z) = \frac{z^2}{z^2(2+a) + z(3+2a) + 2} \quad \text{and} \quad \frac{\partial M}{\partial a} = \frac{-z^2(z^2+2z)}{\{(2+a)z^2 + (3+2a)z + 2\}^2}$$

Therefore

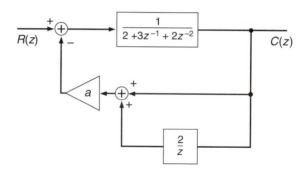

Figure 3.35   Block diagram for Example 3.36.

$$Sa^M = \frac{\partial M}{\partial a}\{\frac{a}{M}\} = \frac{-a(z^2 + 2z)}{z^2(2 + a) + z(3 + 2a) + 2}$$

If a range of frequencies is crucial, we can construct the frequency response. At a value of a = 0.5, the system is stable, so we evaluate the sensitivity function at that value and get the frequency response plot in Fig. 3.36 from the MATLAB statement FREQZ(B,A).

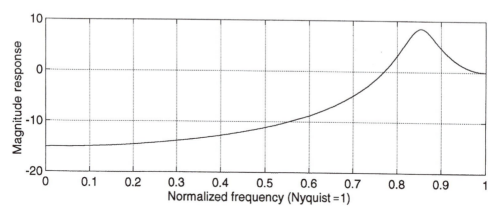

Figure 3.36   Frequency response plot for Example 3.36.

Note that up to about 75% of the sampling frequency the sensitivity function is less than 0 dB. That means that a change in parameter value of a certain percentage will yield that percent change or less in the overall transfer function, as long as the frequency stays less than 75% of the sampling frequency.

3.7 SUMMARY

Sampled-data systems, including ZOH and FOH devices, were discussed in Section 3.2. General-purpose and function-specific DACs and ADCs were covered. Some important criteria in selection of ADCs and DACs such as resolution, dynamic range, conversion time, and sample-and-hold capability were indicated. State variables were introduced in Section 3.3 as a powerful alternative to classical transfer function representation of a system. These methods are used widely with CAD packages and modern control techniques. Dynamical equations of both continuous- and discrete-time systems were shown, and the state-transition matrix was defined as $e^{At} = \phi(t)$, the exponential matrix that appears in the solution to the state equation. We looked briefly at nonlinear systems in Section 3.4 and focused on the use of the MATLAB simulation tool SIMULINK. Stability analysis of discrete-time systems was discussed in Section 3.5. The condition for BIBO stability was shown, and asymptotic stability of the zero-input response of a discrete system was demonstrated. The Routh-Hurwitz stability method was reviewed for continuous systems and an extension of it was considered using the mapping $z = (1 + \dfrac{T}{2}w) \Big/ (1 - \dfrac{T}{2}w)$ for discrete systems. The Jury test was discussed as a quick method of stability checking for a discrete characteristic equation. The Liapunov stability method, which can be used – though not without difficulty – for nonlinear systems, was introduced as a test for asymptotic system stability. Sensitivity of a system transfer function due to a change of parameter was covered in Section 3.6.

PROBLEMS

3.1 Find the discrete-time transfer function C(z)/R(z) for the system in Fig. 3.37.

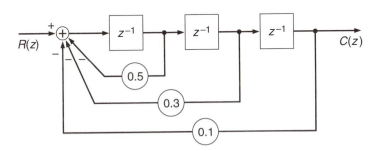

Figure 3.37   System for Problem 3.1.

3.2 Use the C2DM statement from MATLAB or the DISCRETIZ statement from MATRIX$_X$ to obtain the transfer function G(z) for a system that consists of an ideal sampler followed by a ZOH and a transfer function

$$G(s) = \frac{10s + 20}{s^4 + s^3 + 2s^2 + s + 2}$$

Let T = 0.1 second.

3.3 Determine the closed-loop transfer function for the system shown in Fig. 3.38. Assume that T = 0.15 second. Then use MATLAB to plot the unit step response.

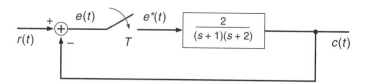

Figure 3.38   Block diagram for Problem 3.3.

3.4 Use MATLAB to determine the unit step response of the system in Fig. 3.39. Let T = 0.05, 0.1, 0.5, and 1.0 second and comment on the difference in responses.

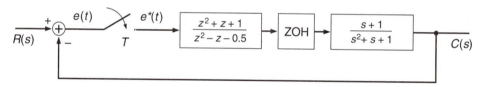

Figure 3.39   Block diagram for Problem 3.4.

3.5 A two-phase control-field AC motor has a transfer function

$$\frac{\theta(s)}{v(s)} = \frac{k}{s(\alpha s + 1)}$$

where α is a function of inertia J and friction f. Cast the system into the following state-variable representations:
   (a) The controllability form
   (b) The observability form
   (c) The Jordan form

3.6 Write the state and output equations for the discrete system represented in Fig. 3.40. Then using the (A,B,C,D) matrices and MATLAB, determine the closed-loop transfer function.

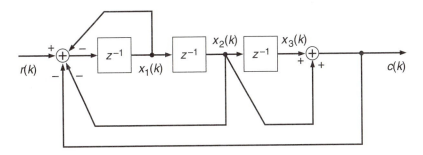

Figure 3.40   Block diagram for Problem 3.6.

3.7 Appendix E defines the state-transition matrix (STM) for continuous-time systems as $e^{At} = \phi(t)$. Show that $\phi(t_2 - t_0) = \phi(t_2 - t_1)\phi(t_1 - t_0)$.

3.8 Determine the state-variable representation for the circuit shown in Fig. 3.41 and find the system's STM.

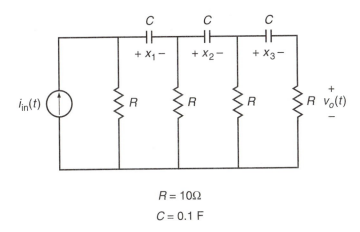

$$R = 10\Omega$$
$$C = 0.1 \text{ F}$$

Figure 3.41   Circuit for Problem 3.8.

3.9 Find the transfer function

$$H(s) = \frac{I_{out}(s)}{V_{in}(s)} = \frac{Y(s)}{R(s)}$$

for Problem 3.8, then cast the circuit into the CF and the OF. Do the state variables in these canonical representations refer to anything physical in the circuit?

3.10 Consider the following continuous-time system:

$$A = \begin{bmatrix} 0 & 0 & -350 \\ 1 & 0 & -155 \\ 0 & 1 & -22 \end{bmatrix}$$

$$B = \begin{bmatrix} 1 \\ 1 \\ 1 \end{bmatrix}$$

$$C = \begin{bmatrix} 0 & 0 & 1 \end{bmatrix}$$

$$D = \begin{bmatrix} 0 \end{bmatrix}$$

(a) Determine the system transfer function.

(b) Determine the system eigenvalues.

(c) Construct the JF representation using the CANON statement from MATLAB.

3.11 Diagonalize the following A matrices using $MATRIX_X$ or MATLAB; then, assuming continuous-time systems, compute the STM in both cases:

$$(a) \ A = \begin{bmatrix} 10 & 20 & 10 \\ 15 & 50 & -15 \\ 20 & 0 & 5 \end{bmatrix}$$

$$(b) \ A = \begin{bmatrix} 2 & 2 & -2 & 2 \\ 1 & 2 & 1 & -1 \\ 1 & 0 & 2 & 0 \\ 1 & 1 & 0 & 2 \end{bmatrix}$$

3.12 Find the STM $\phi(k) = A^k$ for the discrete system

$$A = \begin{bmatrix} -0.2 & 0.2 & 0.5 \\ 0 & 0.2 & 0.3 \\ 0 & 0 & 0.1 \end{bmatrix}$$

$$B = \begin{bmatrix} 1 \\ 1 \\ 2 \end{bmatrix}$$

$$C = \begin{bmatrix} 0 & 1 & 0 \end{bmatrix}$$

$$D = \begin{bmatrix} 0 \end{bmatrix}$$

3.13 For the system in Problem 3.12, determine the unit pulse response.

3.14 Determine the state-transition matrix $\Phi(t)$ for the continuous-time system described by

$$\dot{x}(t) = \begin{bmatrix} 0 & 1 \\ 0 & 10 \end{bmatrix} x(t)$$

Then solve for the state vector $x(t)$ if the initial condition is

$$x(0) = \begin{bmatrix} -2 \\ 2 \\ 1 \end{bmatrix}$$

3.15 Consider the discrete system whose plant model is given by

$$\begin{bmatrix} x_1(k+1) \\ x_2(k+1) \end{bmatrix} = \begin{bmatrix} 0.8 & 0 \\ 0.3 & 2.0 \end{bmatrix} \begin{bmatrix} x_1(k) \\ x_2(k) \end{bmatrix} + \begin{bmatrix} 0.3 \\ 0.8 \end{bmatrix} r(k)$$

$$c(k) = x_1(k) + 2x_2(k)$$

    (a) Find a Jordan form representation for the system.
    (b) Use the z-transform method to find the transfer function
        [C(z)/R(z)].

3.16 Determine the discrete-time dynamical state equation solution for a system with zero initial conditions and input

$$r(k) = \begin{bmatrix} (-1)^k \\ (-0.5)^k \end{bmatrix} u(k)$$

when the A and B matrices are as follows:

$$A = \begin{bmatrix} 0 & 1 \\ 1 & -1 \end{bmatrix} \quad B = \begin{bmatrix} 2 & -1 \\ 0 & -3 \end{bmatrix}$$

3.17 Use the bilinear transformation to transform $T(z) = \dfrac{z + 0.5}{z^2 - z + 0.8}$ to the a form suitable for use with the Routh-Hurwitz criterion and check system stability. Let T=0.2 second.

3.18 Use the Routh-Hurwitz criterion to find the stability of a continuous system that has the characteristic equation

$$q(s) = s^4 + s^3 + 2s^2 + s + 10$$

3.19 Use the Jury test to check stability of a system with the discrete characteristic equation

$$q(z) = z^2 + 0.5z + 5$$

3.20 The characteristic equation of a discrete system is

$$q(z) = z^3 + z^2 + 8z - 2$$

Determine if the system is stable.

3.21 Determine whether the system with the following characteristic equation is stable (using MATRIX$_X$ or MATLAB to get the roots of the polynomial):

$$q(z) = z^5 + z^4 + 8z^3 - 2z^2 - 2z + 1$$

3.22 Use the Jury stability method to determine the range of K needed for the stability of the system in Fig. 3.42.

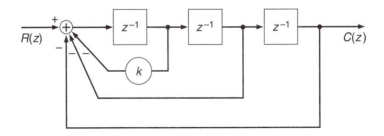

Figure 3.42   System for Problem 3.22.

3.23 Given the matrix

$$A = \begin{bmatrix} 1 & 0 & 0 \\ 1 & 1 & 1 \\ 1 & 1 & 0 \end{bmatrix}$$

of a discrete system, without determining the eigenvalues, use the Liapunov method to check for system stability. Use MATLAB.

3.24 Determine if any of the following matrices are positive definite:

(a) $\begin{bmatrix} 0 & 0 & 0 \\ 0 & 1 & 0 \\ 0 & 0 & 0 \end{bmatrix}$ (b) $\begin{bmatrix} 0 & 0 & 1 \\ 1 & 0 & 1 \\ 0 & 0 & 0 \end{bmatrix}$ (c) $\begin{bmatrix} -1 & -1 \\ -1 & -1 \end{bmatrix}$

3.25 Use the Liapunov method to find the stability of the following discrete system. In this problem use MATLAB as a check on results from hand calculations.

$$x(k+1) = \begin{bmatrix} 1 & -1 \\ 0 & 1 \end{bmatrix} x(k)$$

3.26 Linearize the following discrete system and check it for stability in the neighborhood of its stationarity points:

$$x_1(k+1) = 2x_1(k)x_2(k)^2 - x_1(k)^2$$
$$x_2(k+1) = 2x_1(k)x_2(k)$$

3.27 Determine the sensitivity to the change of parameter k of the closed-loop transfer function of the system shown in Fig. 3.43.

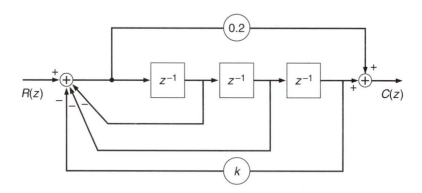

Figure 3.43   System for Problem 3.27.

REFERENCES

1. C. L. Philips and H. T. Nagle, *Digital Control Systems Analysis and Design,* 2nd ed., Prentice Hall, Upper Saddle River, NJ 1990.

2. H. F. Van Landingham, *Introduction to Digital Control Systems,* Macmillan, New York, 1985.

3. G. F. Franklin, J. D. Powell, and M. L. Workman, *Digital Control of Dynamic Systems,* 2nd ed, Addison-Wesley, Reading, MA, 1990.

4. R. C. Dorf, *Modern Control Systems,* Addison-Wesley, Reading, MA, 1980.

5. C. T. Chen, *Linear System Theory and Design,* CBS College Publishing, New York, 1984.

6. *Data Converter Reference Manual,* Vol. I, Analog Devices, Inc., Norwood, MA, 1992.

7. *Data Converter Reference Manual,* Vol. II, Analog Devices, Inc., Norwood, MA, 1992.

8. *ADSP-2100 Family User's Manual,* Analog Devices, Inc., Norwood, MA, 1993.

9. *Test and Measurement Catalog,* Hewlett-Packard, Palo Alto, CA, 1991.

10. P. T. Cook, *Nonlinear Dynamical Systems,* Prentice Hall International, London, 1986.

Design of Digital Control Systems

4.1 INTRODUCTION

In Chapter 3 we concentrated on some of the main topics in the analysis of digital systems. Analysis is a discipline in its own right, but from a broader perspective, analysis is considered to be part of design. That is, design always includes more than analysis. In particular, design specifications, which are not at issue in analysis, play a crucial role in the design procedure. In this chapter we study several of the ideas associated with the design of digital control systems from the classical point of view. Design from the modern state variable point of view is taken up in Chapter 6. In the chapter in-between, Chapter 5, we detail the use of DSPs and show how they can be configured to function as controllers in discrete systems. The primary aim of the design procedure, whether classical or modern, is to generate a mathematical description of a controller which can then be implemented in several ways, for example, as a DSP.

Control engineers sometimes design digital controllers by first designing them in the analog domain using familiar analog techniques. Then they use transformations, such as the bilinear transformation, to transform the analog controllers into the digital domain. This design methodology is called *indirect design*. An advantage of indirect design is that most designers are familiar with analog design techniques such as Bode plots, root-locus, and the Routh-Hurwitz method. Another advantage is that designers can use familiar continuous domain design specifications. This methodology's major drawback is that it ignores samplers and hold devices, which play a crucial role in sampled-data systems. For sampled-data systems with high sampling rates, the results of indirect design are generally acceptable.

With lower sampling rates the results are less acceptable. There is an alternative to indirect design, good for low and high sampling rates, which is to design directly in the z-domain. This generally preferred method is known as *direct design*. Herein a given continuous plant expressed in analog form is transformed to the z-domain by any of several methods discussed in Chapter 3, for instance, the zero-order-hold method. Then z-domain transfer functions are manipulated. In direct design, several techniques are available to achieve the desired system performance. Several of the classical analog techniques, in fact, map over to the z-plane with very little modification. Fig. 4.1 shows the design flow for both direct and indirect design methodologies.

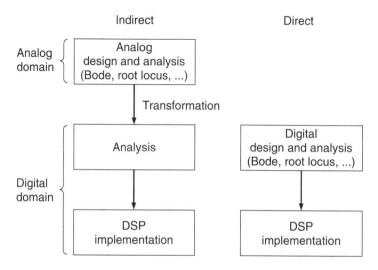

Figure 4.1   Direct and indirect design flows in control systems.

In Section 4.2 we review dynamic response and steady-state parameters from classical control theory for continuous-time systems. Then we investigate the translation of these ideas into digital control systems. In general, steady-state behavior is a function of the numerator dynamics of the closed-loop transfer function. Also, in general, transient response behavior (rise time, settling time, and so on) is a function of the denominator of the closed-loop transfer function.

In Section 4.3 we review two classical design and analysis tools: Bode plots and root-locus method. The methods of construction for these are the same for the z-plane as for the s-domain. Only the interpretation of design and analysis results is different.

Section 4.4 covers the notion of compensation in control systems. Compensators add poles or zeros to the open-loop transfer function in order to obtain desired closed-loop responses. Bode plots are used to study phase lead and phase lag compensators. Proportional-integral-derivative (PID) and deadbeat controllers are introduced.

As mentioned, we take up with additional design procedures emphasizing modern techniques in Chapter 6. It should be stressed that modern techniques have not usurped the role of classical design techniques. Both are widely used in industry. Modern approaches are more sophisticated and can be used for more complex systems, but for most lower-order systems the classical procedures are entirely adequate and often preferred. Many designers feel that classical frequency-domain manipulation of Bode plots or root-locus trajectories is a more concrete activity than the abstract manipulation associated with modern state variable techniques. Concreteness keeps them more connected to the real physical systems they are designing.

4.2 CONTROL SYSTEM DESIGN PARAMETERS

Dynamic Response Parameters

Design parameters specify desired control system designs. These parameters specify such things as dynamic response, the ability of a system to keep the error negligible as input changes, steady-state accuracy, and system stability. Before we explore these ideas in the realm of discrete systems, let us consider the continuous system of Fig. 4.2 in canonical form.

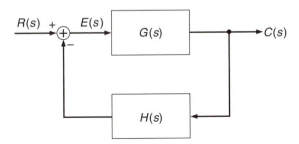

Figure 4.2   Canonical continuous system.

As we know, the overall transfer function of the system is

$$T(s) = \frac{C(s)}{R(s)} = \frac{G(s)H(s)}{1 + G(s)H(s)} \qquad (4.1)$$

We will look in detail at the case where the system in Fig. 4.2 is a second-order system that has the generalized second-order transfer function of the form

$$T(s) = \frac{\omega^2_n}{s^2 + 2\xi\omega_n s + \omega^2_n} \tag{4.2}$$

where ω_n is the undamped natural frequency of the system and ξ is the damping ratio. Many systems are expressible in this form or can be approximated by second-order systems or are composed of cascaded combinations of such systems. This standard form can arise out of a system with unity gain feedback [H(s) = 1] and an open-loop transfer function of the form:

$$G(s) = \frac{\omega^2_n}{s(s + 2\xi\omega_n)} \tag{4.3}$$

We want to investigate the dynamic response of the generalized second-order system to a step input. When we apply a unit step input $R(s) = \frac{1}{s}$, the output of the system is

$$C(s) = \frac{\omega^2_n}{s(s^2 + 2\xi\omega_n s + \omega^2_n)} \tag{4.4}$$

The Laplace transform of the output is from Appendix C:

$$c(t) = 1 - \frac{1}{\sqrt{1-\xi^2}} e^{-\xi\omega_n t} \sin(\omega_n \sqrt{1-\xi^2}\, t + \theta) \tag{4.5}$$

where

$$\theta = \tan^{-1} \frac{\sqrt{1-\xi^2}}{\xi} \tag{4.6}$$

Fig. 4.3 shows different types of responses for a step input as a function of the damping ratio ξ. When $\xi > 1$, the response is *overdamped* and shows no overshoot. When $\xi < 1$, the response indicates oscillatory behavior that gradually dies out and is said to be *underdamped*. A value of $\xi = 1$ yields a *critically damped* system that responds as fast as possible without exhibiting any overshoot.

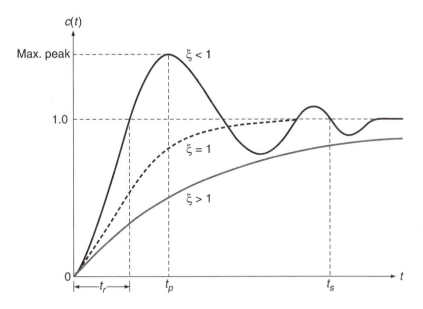

Figure 4.3 Dynamic response of a second-order system for a step input.

When $\xi > 1$, the response is *overdamped* and shows no overshoot. When $\xi < 1$, the response indicates oscillatory behavior that gradually dies out and is said to be *underdamped*. A value of $\xi = 1$ yields a *critically damped* system that responds as fast as possible without exhibiting any overshoot.

We can now define the following dynamic response parameters based on Fig. 4.3:

- *Peak time* (t_p): time required for peak overshoot:

$$t_p = \frac{\pi}{\omega_n \sqrt{1-\xi^2}} \tag{4.7}$$

- *Percent overshoot* (P.O.): value of the output response at peak time t_p:

$$\text{P.O.} = 100\exp(\frac{-\xi\pi}{\sqrt{1-\xi^2}}) \tag{4.8}$$

- *Settling time* (t_s): time the response requires to settle within the desired percentage of its final output. For a second-order system, the response is usually within 2% after it reaches the settling time:

$$t_s \approx \frac{4}{\xi \omega_n} \qquad (4.9)$$

- *Rise time* (t_r): time the response requires to go from 10% to 90% of its final value:

$$t_r = \frac{0.8 + 2.5\xi}{\omega_n} \qquad (4.10)$$

Example 4.1

Determine the poles of T(s) from Equation 4.2 if we want a settling time of 4 seconds and a peak time of 1 second for a unit step input. Then determine the resulting percent overshoot.

Solution: From Equation 4.9

$$4 = \frac{4}{\xi \omega_n} \Rightarrow \xi \omega_n = 1$$

and from Equation 4.7

$$1 = \frac{\pi}{\omega_n \sqrt{1 - \xi^2}}$$

from which

$$\omega_n^2 (1 - \xi^2) = \pi^2 = 9.8696 = \frac{1 - \xi^2}{\xi^2}$$

which implies that $\xi = 0.303$ and $\omega_n = 3.3\,\text{rad}\,/\,\text{sec}$. Then the percent overshoot from Equation 4.8 is

$$\text{P.O.} = 100\exp(\frac{-\xi\pi}{\sqrt{1 - \xi^2}}) = 100e^{-1.086} = 33.75\%$$

Example 4.2

Fig. 4.4 shows an op-amp circuit with feedback that can be used as a low-pass filter. Determine the transfer function

$$T(s) = \frac{V_o(s)}{V_i(s)}$$

assuming an ideal op amp and calculate the undamped natural frequency and the damping ratio. What are the rise time and peak time of the circuit if we let $v_i(t) = u(t)$ volts?

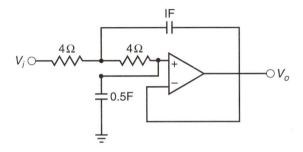

Figure 4.4   Op-amp circuit for Example 4.2.

Solution: Letting $V(s)$ be the node voltage at the node between the two 4Ω resistors, we can write the node equations

$$[V_i(s) - V(s)]\frac{1}{4} = \frac{V(s) - V_o(s)}{1/s} + \frac{V(s) - V_o(s)}{4}$$

$$\frac{V(s) - V_o(s)}{4} = \frac{V_o(s)}{1/0.5s}$$

eliminating $V(s)$ and solving for $T(s)$, we arrive at

$$T(s) = \frac{V_o(s)}{V_i(s)} = \frac{0.125}{s^2 + 0.5s + 0.125}$$

which implies that

$$\omega_n = 0.345 \text{rad} / \sec$$
$$\xi = 0.707$$

Then we solve for the rise and peak times using Equations 4.10 and 4.7:

$$t_r = \frac{0.8 + 2.5\xi}{\omega_n} = 7.25 \text{ seconds}$$

$$t_p = \frac{\pi}{\omega_n \sqrt{1 - \xi^2}} = 12.56 \text{ seconds}$$

We normally translate these dynamic time response specifications for continuous systems into information about pole locations in the s-plane. For example, referring to Fig. 4.5, a constant ω_n is indicated by a circle in the s-plane with radius ω_n and poles of T(s) are indicated in complex-conjugate pairs at points along the circle's circumference. Of particular interest are poles along the constant ω_n locus in the second and third quadrants of the s-plane. They correspond to ξ values between zero and 1. A value of $\xi = 0$, for instance, implies poles of T(s) at $s = \pm j\omega_n$ which is the oscillatory case, while $\xi = 1$ implies repeated poles at $s = -\omega_n$, which is the critically damped case. Poles along a constant ξ-locus in the s- plane will vary radially from the origin as $\omega_n \rightarrow \infty$.

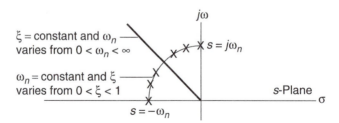

Figure 4.5   Pole variations in the s-plane as ξ and ω_n vary.

In Chapter 2 we examined the transformation of different regions of the s-plane into the z-plane according to the mapping rule $z = e^{sT}$. To handle the transformation of transfer functions, for example T(s) into T (z), several techniques were indicated in Chapter 2. All of these techniques employ different approximations to $z = e^{sT}$. One of the most popular techniques, illustrated in the next example, is the bilinear transformation.

Example 4.3[1]

The following analog transfer function is used in a DC motor, two-axis gimbaled platform system for a rate-loop second-order compensator:

$$G(s) = 1000 \frac{s^2 + 68.2s + 3943}{s^2 + 2512s + 6.31 \times 10^6}$$

We can find the digital equivalent of the analog transfer function above by using the bilinear transformation with frequency prewarping. The sampling frequency to be used in converting to a digital equivalent is $f = 4020$ Hz. So the sampling period is $T = 248.76$ μsec. The characteristic equation of this analog transfer function is

$$s^2 + 2512s + 6.31 \times 10^6 = 0$$

which fits the standard, second-order form

$$s^2 + 2\xi\omega_n s + \omega^2_n = 0$$

The natural frequency $\omega_n = 2511.97$ rad/s. To compensate for nonlinear mapping of analog-to-digital frequencies by the bilinear transformation method, the natural frequency is prewarped according to Equation 2.92:

$$\omega_p = \frac{2}{T} \tan \frac{\omega_0 T}{2} = \frac{2}{248.76 \times 10^{-6}} \tan \frac{2511.9713 \times 248.76 \times 10^{-6}}{2} = 2597.03 \text{ rad / s}$$

The ω_0 in this equation is the natural frequency ω_n. This prewarping scheme matches exactly the natural frequency in the analog and digital domains for the compensator. To obtain the prewarped version of the analog transfer function, the complex variable s in the original transfer function is replaced with $(\omega_0/\omega_p)s$. It is therefore convenient to compute the ratio

$$\frac{\omega_0}{\omega_p} = \frac{2511.9713}{2597.03} = 0.9672$$

The prewarped G(s), i.e., $Q(s)$ is then computed as

---

[1] Portions reprinted by permission of Texas Instruments. Reprinted with permission from authors. See Reference [9].

$$Q(s) = 1000\frac{(0.9672s)^2 + 68.2(0.9672s) + 3943}{(0.9672s)^2 + 2512(0.9672s) + 6.31 \times 10^6}$$

$$= 1000\frac{s^2 + 70.51s + 4214.87}{s^2 + 2597.16s + 6.75 \times 10^6}$$

Bilinear transformation is next applied to $Q(s)$. The continuous variable s is replaced by the expression that involves the discrete variable z:

$$s = \frac{2}{T}\frac{z-1}{z+1}$$

This produces the discrete transfer function D (z).
For the compensator,

$$D(z) = Q(s)\Big|_{s = \frac{2(z-1)}{T(z+1)}}$$

$$= 1000\frac{s^2 + 70.51s + 4214.87}{s^2 + 2597.16s + 6.75 \times 10^6}\Big|_{s = \frac{2(z-1)}{248.76 \times 10^{-6}(z+1)}}$$

After further computations,

$$D(z) = 706.76\frac{1.0 - 1.9824z^{-1} + 0.9826z^{-2}}{1.0 - 1.2548z^{-1} + 0.5474z^{-2}}$$

The final step is the gain adjustment in the digital transfer function. This can be accomplished by matching the analog and digital gains at some predetermined frequency, for example, dc.
For the dc case, $s = j\omega = 0$, and from the bilinear transformation:

$$z = \frac{2 + sT}{2 - sT} = 1$$

Therefore, at dc, Q (0) = D(1). For this transfer function, Q (0) = 0.6249, D (1) = 0.5072. If Q (0) = K x D (1), then the constant K becomes

$$\frac{0.6249}{0.5072} = 1.2321$$

The final form of the digital equivalent transfer function is

$$D(z) = 870.77 \frac{1.0 - 1.9824z^{-1} + 0.9826z^{-2}}{1.0 - 1.2548z^{-1} + 0.5474z^{-2}}$$

where the gain of 870.77 is the product of $K \times$ (the unadjusted digital gain), i.e., 870.77 = 1.2321 x 706.76.

If we translate the poles of

$$T(s) = \frac{\omega^2_n}{s^2 + 2\xi\omega_n s + \omega^2_n}$$

with the mappings $z = e^{sT}$ and $s = (1/T)\ln z$, then a locus of pole movements for constant ξ (a radial variation in the s-plane) becomes a logarithmic spiral in the z-plane (see Fig. 2.24). Also a locus of constant ω_n (a circle in the s-plane) becomes a line drawn at right angles to a spiral of a constant ξ. In the z-plane, in addition to ξ and ω_n being adjustable, the sampling period T is also an adjustable parameter and its value will influence pole locations. The essential point here is that using the indicated mappings, the LHP poles of T(s) will map into poles of T(z) inside the unit circle in the z-plane.

However, it is not realistic to translate T(s) directly to T(z) using $s = (1/T)\ln z$ because this will transform finite polynomials into infinite polynomials. Also, in a real system we have samplers and zero- or higher-order hold devices to contend with. A more typical situation, illustrated in the next example, would be one wherein we have a $G(s) = \omega^2_n \Big/ s(s + 2\xi\omega_n)$ transfer function preceded by a sampler and a ZOH device in the forward loop and a unity gain feedback. We will see that even though poles in the LHP may indicate good transient response with modest overshoot for the closed-loop continuous system, upon translating to the z-plane – depending on T – we can get poles that indicate excessive overshoot and even instability for the discrete system.

Example 4.4

Consider the system of Fig. 4.6 where

$$G_p(s) = \frac{\omega_n^2}{s(s + 2\xi\omega_n)}$$

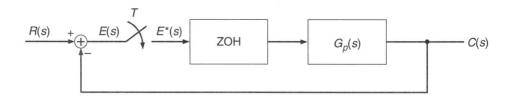

Figure 4.6   Sampled second-order system.

Let $\xi = 0.5$ and $\omega_n = 1.0$ rad/s so the step response of the continuous system would be slightly underdamped. Now, for the sampled system

$$G(s) = G_{ZOH}(s)G_p(s) = \left(\frac{1 - e^{-sT}}{s}\right)\frac{\omega_n^2}{s(s + 2\xi\omega_n)}$$

$$G(z) = (1 - z^{-1})Z\left\{\frac{\omega_n^2}{s^2(s + 2\xi\omega_n)}\right\}$$

and to get the z-transform of the term in braces we must first invert the Laplace transform to get f(t). Then we discretize f(t) to get f(kT) and take the z-transform of f(kT).

$$\frac{\omega_n^2}{s^2(s + 2\xi\omega_n)} = \frac{1}{s^2(s + 1)} = \frac{A}{s} + \frac{B}{s^2} + \frac{C}{s + 1} = \frac{-1}{s} + \frac{1}{s^2} + \frac{1}{s + 1}$$

Thus

$$f(t) = -u(t) + tu(t) + e^{-t}u(t)$$

$$f(kT) = -u(kT) + kTu(kT) + e^{-kT}u(kT)$$

Then the z transform becomes

$$G(z) = (1 - z^{-1})\{\frac{-z}{z+1} + \frac{Tz}{(z-1)^2} + \frac{z}{z - e^{-T}}\}$$

Considering the case of T = 1 second, we get

$$G(z) = -1 + \frac{1}{z-1} + \frac{z-1}{z-0.367} = \frac{0.369z + 0.262}{z^2 - 1.369z + 0.369}$$

But we really need T(z), the closed-loop transfer function, which is

$$T(z) = \frac{G(z)}{1 + G(z)} = \frac{0.369z + 0.262}{z^2 - z + 0.631}$$

Using the Jury test and since the constant term in the characteristic polynomial is <1, we check $\Delta(z)$ at z = 1 and -1:

$$\Delta(1) = 1 - 1 + 0.631 > 0$$
$$\Delta(-1) = 1 + 1 + 0.631 > 0$$

and the discrete system is stable.

Now look at other T values. There is a MATLAB expression to convert a continuous system to discrete form:

[NUMd, DENd]=C2DM (NUM, DEN, Ts,'method')

All we need to do is order the numerator and denominator of the given G(s), let the sampling rate be Ts in seconds, and specify the method of transformation. We use the ZOH method in this example. The following statements return G(z) for the case of T = 0.1 second:

NUM=[1]

DEN=[1 1 0]

Ts=0.1

[NUMd, DENd]=C2DM (NUM, DEN, Ts,'zoh')

and we get

$$G(z) = \frac{0.0048z + 0.0047}{z^2 - 1.9048z + 0.9048}$$

from which the closed-loop transfer function becomes

$$T(z) = \frac{0.0048z + 0.0047}{z^2 - 1.9000z + 0.9095}$$

The Jury test again indicates a stable system. Generally, the smaller the Ts value, the closer the discrete system approximates the original continuous system. Increasing Ts, however, can produce problems. For Ts = 2 we get

$$G(z) = \frac{1.1353z + 0.5940}{z^2 - 1.1353z + 0.1353}$$

and

$$T(z) = \frac{1.1353z + 0.5940}{z^2 + 0.7293}$$

which is also stable. But Ts = 5 yields

$$G(z) = \frac{4.0067z + 0.9596}{z^2 - 1.0067z + 0.0067}$$

and

$$T(z) = \frac{4.0067z + 0.9596}{z^2 + 3z + 0.9663}$$

and T(z) fails the Jury test, indicating poles outside the unit circle and an unstable system. One more MATLAB statement is useful for this example:

DSTEP (NUMd, DENd)

Entering the numerator and denominator of T(z), this statement will produce the discrete unit step response of the closed-loop system. We present these plots in Fig. 4.7, 4.8, 4.9, and 4.10 for the cases of Ts = 0.1, 1.0, 2.0, 5.0 respectively. Note that for Ts = 0.1 the step response is similar to the expected continuous result. For Ts=5, on the other hand, the response becomes rapidly unbounded.

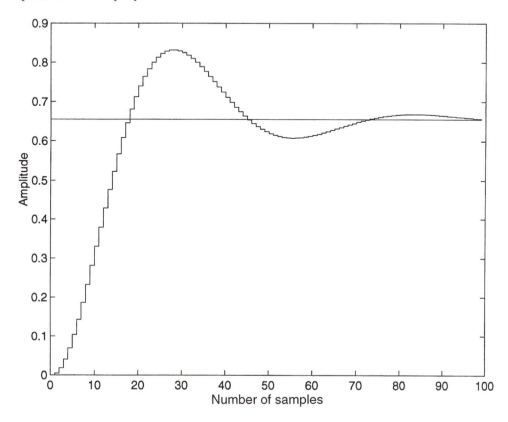

Figure 4.7 Plot for Ts=0.1 second.

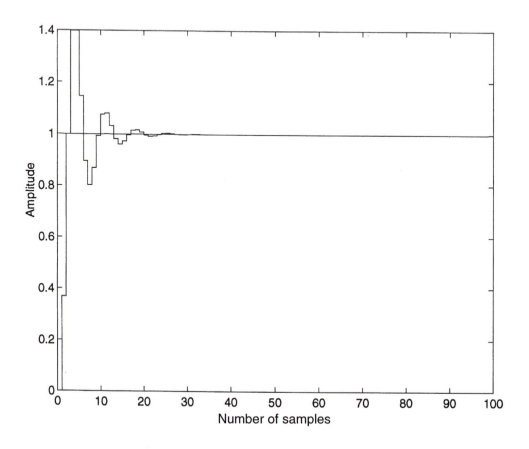

Figure 4.8   Plot for Ts=1 second.

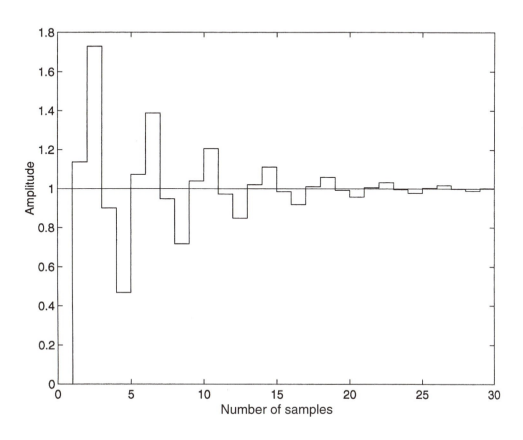

Figure 4.9   Plot for Ts=2 seconds.

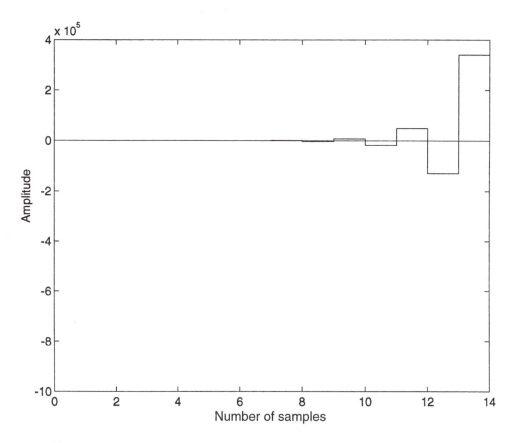

Figure 4.10   Plot for Ts=5 seconds.

Steady-State Parameters

It is often the case that we want control systems to follow or track inputs, like step or ramp functions, as quickly as possible. The difference between input and output, the system error function, is of fundamental concern to control-system designers. Based on the notion of steady-state error, several useful results can be developed. Consider the unity gain feedback system in Fig. 4.11.

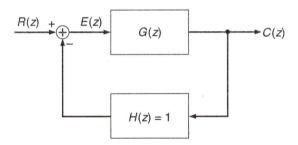

Figure 4.11 Unity gain feedback discrete system.

The transfer function and the error signal of the system are

$$T(z) = \frac{G(z)H(z)}{1 + G(z)H(z)} = \frac{G(z)}{1 + G(z)} \tag{4.11}$$

and

$$E(z) = R(z) - C(z) = \frac{R(z)}{1 + G(z)} \tag{4.12}$$

If we apply the unit-step input $R(z) = \dfrac{1}{\left(1 - z^{-1}\right)}$, the steady state error of the system due to the unit step input is

$$e_{ss} = \lim_{k \to \infty} e(k) \tag{4.13}$$

But the final value theorem allows us to write

$$\lim_{k \to \infty} e(k) = \lim_{z \to 1} (1 - z^{-1})E(z) \tag{4.14}$$

Recall that the final value theorem required that the closed-loop system under consideration be stable. This restriction should make sense because for an unstable system the error would grow without bound and e_{ss} would be infinite. The expression in Equation 4.13 becomes

$$\lim_{z \to 1}(1 - z^{-1})(\frac{1}{1 - z^{-1}})(\frac{1}{1 + G(z)}) = \lim_{z \to 1}\frac{1}{1 + G(z)} = \frac{1}{1 + G(1)} \qquad (4.15)$$

and if we let

$$K_p = \lim_{z \to 1} G(z) = G(1) \qquad (4.16)$$

we can write

$$e_{ss} = \frac{1}{1 + K_p} \qquad (4.17)$$

System Type is an important idea for discrete as well as continuous systems. A type 0 discrete system has G (z) with no poles at z = 1, a type 1 system has a G (z) with one pole at z = 1, and so on. Thus for a type 0 system with unit step input, G (1) is finite and e_{ss} is available from Equation 4.17, but for type 1 or higher system, G (1) is infinite and e_{ss} is zero.

Now, when we apply the unit ramp input $\left[R(z) = Tz \big/ (z-1)^2 \right]$ to the system of Fig. 4.7, we get the error

$$E(z) = R(z)\frac{1}{1 + G(z)} = \frac{Tz}{(z - 1)^2}\frac{1}{1 + G(z)} \qquad (4.18)$$

Then

$$e_{ss} = \lim_{z \to 1}(1 - z^{-1})E(z) = \frac{T}{\lim_{z \to 1}(z - 1)G(z)} \qquad (4.19)$$

and if we let

$$k_v = \frac{1}{T}\lim_{z \to 1}(z - 1)G(z) \qquad (4.20)$$

we can write

$$e_{ss} = \frac{1}{K_v} \qquad (4.21)$$

Equation 4.21 is useful for the case when G(z) is a type 1 system. If it is type 0, e_{ss} is infinite because K_v is zero. If type 2 or above, K_v is infinite and e_{ss} is zero.

When we apply a unit parabolic input

$$R(z) = \frac{T^2 z(z+1)}{2(z-1)^3} \qquad (4.22)$$

we get the error

$$E(z) = \frac{T^2 z(z+1)}{2(z-1)^2} \frac{1}{1+G(z)} \qquad (4.23)$$

and

$$e_{ss} = \lim_{z \to 1} E(z) = \lim_{z \to 1} \frac{T^2}{(z-1)^2} \frac{1}{(1+G(z))} \qquad (4.24)$$

Letting

$$K_a = \frac{1}{T^2} \lim_{z \to 1} (z-1)^2 G(z) \qquad (4.25)$$

we can write

$$e_{ss} = \frac{1}{K_a} \qquad (4.26)$$

But this assumes that G(z) is a type 2 system. If it is a type 1 or 0 system, e_{ss} is infinite because K_a is zero, and for type 3 systems or above, K_a is infinite and e_{ss} is zero. We can summarize these results in Table 4.1, where the elements of the matrix are e_{ss} values.

Table 4.1   Steady-State Errors

	System Type	0	1	2	3
Input					
Step		$\dfrac{1}{1+K_p}$	0	0	0
Ramp		∞	$\dfrac{1}{K_v}$	0	0
Parabolic		∞	∞	$\dfrac{1}{K_a}$	0

In conventional servo-control systems, the step, ramp and parabolic inputs are typically referred to as the position, velocity, and acceleration inputs respectively. Basic design questions are questions such as the following: If the input is a position input and we want an e_{ss} value of 0.1 or less, what type system do we need and what is the value of some parameter in the system selected? These kinds of specifications are often combined with other specifications that deal with stability of the system. Stability issues in design will be taken up shortly.

Example 4.5

For the system of Fig. 4.12, we can find the velocity error constant K_v, using Equation 4.20. For

$$G(z) = \frac{z}{(z-0.2)(z-1)}$$

which indicates a type 1 system, K_v becomes

$$K_v = \frac{1}{T}\lim_{z \to 1}(z-1)\frac{z}{(z-0.2)(z-1)} = \frac{1.25}{T}$$

and from Table 4.1 we see that $e_{ss} = \dfrac{1}{K_v} = \dfrac{T}{1.25}$. Therefore, if we desire a steady-state error of less than 0.01 in magnitude, we will need $T \leq 0.0125$ second or a sampling frequency of at least 80 Hz.

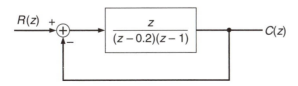

Figure 4.12 System for Example 4.5.

Example 4.6

Consider the system on Fig. 4.13. Determine the range of α and the sampling period T over which the system is both stable and has a steady-state error of 0.01 for acceleration inputs.

Solution:

$$G(z) = \frac{2z^2 + \alpha}{(z-1)^2}$$

indicates a type 2 system and from Table 4.1

$$e_{ss} = \frac{1}{K_a}$$

But

$$K_a = \frac{1}{T^2} \lim_{z \to 1} (z-1)^2 G(z) = \frac{1}{T^2}(2+\alpha)$$

and

$$e_{ss} = 0.01 = \frac{1}{K_a} = \frac{T^2}{(2+\alpha)} = \frac{1}{100} \quad \text{or} \quad (2+\alpha) = (10T)^2$$

Let us look first at the range of α for stability.

$$T(z) = \frac{G(z)}{1+G(z)} = \frac{\dfrac{2z^2+\alpha}{(z-1)^2}}{1+\dfrac{2z^2+\alpha}{(z-1)^2}} = \frac{2z^2+\alpha}{3z^2-2z+(\alpha+1)}$$

and applying the Jury test yields

$$F(z) = 3z^2 - 2z + (\alpha+1)$$

and we need

$$F(1) > 0$$

$$F(-1) > 0$$

$$|a_0| < a_2$$

where

$$a_0 = (\alpha+1)$$
$$a_1 = -2$$
$$a_2 = 3$$

$F(1) > 0$ indicates that $\alpha + 2 > 0$, $F(-1) > 0$ indicates that $\alpha + 6 > 0$, and $|a_0| < a_2$ indicates that $-4 < \alpha < 2$. The range of stability then is $-2 < \alpha < 2$. Now to find the appropriate range of T, refer to the constraint equation derived above: $(2+\alpha) = (10T)^2$. As α ranges from -2 to 2, T ranges from 0 to 0.2 second. Therefore, as long as α is in its stable range, a value of T less than 0.2 second will meet the design specs.

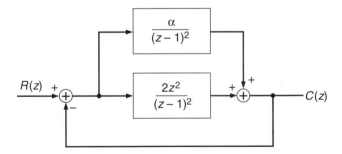

Figure 4.13   System for Example 4.6.

4.3 CONVENTIONAL DESIGN TOOLS

In this section we consider two classical design tools, root-locus and Bode plots, and apply them to discrete systems. The reader probably already knows how to construct root-locus and Bode plots for analog systems from a study of classical control systems. Root-locus construction techniques in the z-domain are identical to those in the s-domain, but the interpretation of results is different in the two planes. Bode plots are used as stability indicators as well as design correction tools in both the analog and discrete cases. CAE software packages such as MATRIX$_X$ and MATLAB are used extensively to generate root-locus and Bode plots of both continuous and discrete systems (see Appendix A).

Root-Locus Method

The root-locus method is a graphical method used to obtain the locus of roots (poles) of the closed-loop system when a parameter changes. The root-locus design technique allows us to move the roots of the characteristic equation into desired locations in the z-plane (or s-plane if analog systems are at issue) by selecting appropriate parameter values or by adding compensator poles and/or zeros. The main advantage of this method is that we need only the positions of the poles and zeros of the open-loop transfer function to obtain the locus of closed-loop roots. To use the root-locus method, complete the following steps:

1.   Write the characteristic equation as

$$q(z) = 1 + f(z) = 0 \qquad (4.27)$$

then rearrange the equation, if necessary, so that the parameter of interest, K, appears as the multiplying factor in the form

$$1 + KP(z) = 0 \tag{4.28}$$

2. Factor $P(z)$, if necessary, then write the polynomial in the form of poles and zeros:

$$1 + K \frac{\prod_{i=1}^{m}(z + z_i)}{\prod_{j=1}^{n}(z + p_i)} = 0 \tag{4.29}$$

   then locate the poles and zeros in the z-plane.

3. The locus of the roots of the characteristic equation $1 + KP(z) = 0$ starts at the poles of $P(z)$ and ends at the zeros of $P(z)$ as K increases from zero to infinity.

4. Any part of the root locus on the real axis of the z-plane always lies on a section of the real axis to the left of an odd number of poles and zeros.

5. The N sections of loci must end at zeros at infinity, where $N = n_p - n_z$, the number of finite poles minus the number of finite zeros. These sections go to zeros at infinity along asymptotes as K approaches infinity. The center of the asymptotes is a point on the real axis given by

$$\delta = \frac{\sum \text{poles} - \sum \text{zeros}}{n_p - n_z} \tag{4.30}$$

   and the angle of the asymptotes with respect to the real axis is

$$A = \frac{(2b + 1)180^{\circ}}{n_p - n_z} \tag{4.31}$$

   where $b = 0, 1, \ldots, (n_p - n_z - 1)$

6. The root locus leaves the real axis at a point called the *breakaway point*. At this breakaway point we have a multiplicity of roots. The breakaway point occurs where K is maximum on the real axis. But this is the same point where $-P(z)$ is minimum or $\dfrac{dP(z)}{dz} = 0$.

7. If the root locus crosses the unit circle in the z-plane, we can find the exact location of the crossing point using the Jury test.

Example 4.7

Consider the discrete open-loop transfer function

$$G(z) = \frac{Kz^2}{(z-1)(z-0.9)}$$

for a unity gain feedback system. By applying the procedure above to G(z), we obtain the root locus as in Fig. 4.14. The arrow on the locus shows the direction of increasing gain K.

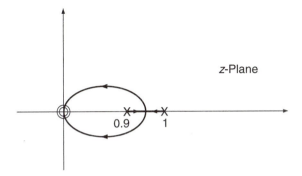

Figure 4.14   Root locus of Example 4.7.

If we desire a system with closed-loop double poles, we could calculate the breakaway point as that point where

$$\frac{dP(z)}{dz} = 0 \text{ or } \frac{d\dfrac{z^2}{(z-1)(z-0.9)}}{dz} = 0$$

From this we get the polynomial:

$$2(z^2 - 1.9z + 0.9) = z^2 - 0.9z + z^2 - z \text{ or } 1.8 = 1.9z$$

Thus the breakaway point occurs at z = 0.947 and the value of K at that point is

$$K = \frac{-1}{P(z)} = \frac{-1}{\dfrac{0.947^2}{(0.947-1)(0.947-0.9)}} = 0.00276$$

With this value of K the closed-loop system has a double pole at $z = 0.947$.

Most readers are no doubt aware of CAE tools for doing continuous-time root-locus calculations. MATLAB provides the RLOCUS statement to calculate and plot a root locus for a given numerator and denominator of a P(z) function. The same routine can be used for both continuous and discrete plots. All we need to do is enter NUM and DEN as vectors of coefficients in descending order of the given P(z) function. For Example 4.7, for instance, the MATLAB plot appears as in Fig. 4.15.

If we are interested in determining a K value associated with particular roots, we can write R=RLOCUS (NUM, DEN, K) and specify the desired range of K. MATLAB will produce the root-locus plot and print out closed-loop pole values and their associated K value. A little trial and error should convince the reader that the $z = 0.947$ and $K = 0.00276$ values are correct for Example 4.7. The RLOCFIND statement is also useful in this regard and will be used in the next example.

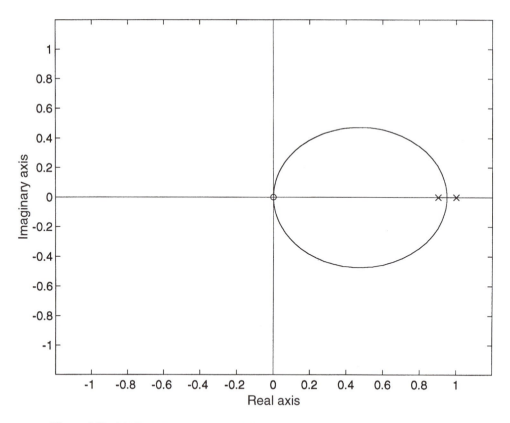

Figure 4.15   MATLAB plot for Example 4.7.

For a second-order discrete system, it is generally desired that the poles of the closed-loop system be in the first and fourth quadrants of the interior of the unit circle. For higher-order systems we often have a pair of dominant poles that we would like to be in the first and fourth quadrants of the interior of the unit circle.

Example 4.8

Consider the system with characteristic equation

$$1 + KGH(z) = 1 + K\frac{(z - 0.5)(z - 0.3)(z - 1)}{z(z + 1)(z + 2)} = 0$$

We want to plot the root locus of KGH (z). To avoid having to multiply out the numerator and denominator to get NUM and DEN as vectors, we can employ another MATLAB statement: ZP2TF. This statement allows us to work with given poles and zeros and essentially does the multiplication for us. We can write

    Z=[0.5 0.3 1.0]
    P=[0.0 -1.0 -2.0]
    [NUM, DEN]=ZP2TF (Z, P, 1)
    RLOCUS (NUM, DEN)

and the root locus is plotted as in Fig. 4.16. (The "1" in the ZP2TF expression is the transfer function gain which we nominally assume to be "1" and then the root locus varies the gain from zero to infinity.)

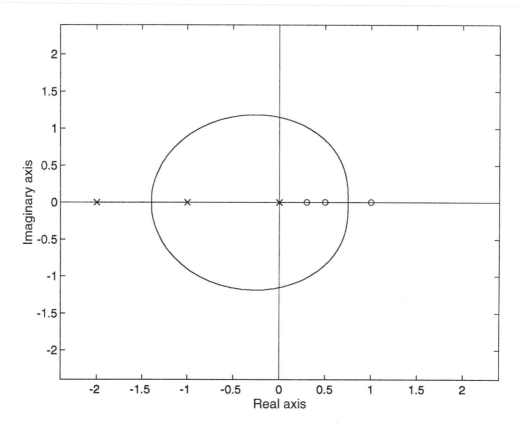

Figure 4.16   Root locus for Example 4.8.

Obviously, the system is unstable if K is too small. We can calculate how big it needs to be to have poles all inside the unit circle by using the Jury test. The characteristic equation becomes

$$(K+1)z^3 + (3-1.8\,K)z^2 + (2+0.95K)z - 0.15K = 0.$$

The Jury array is

$$
\begin{array}{cccc}
K+1 & 3-1.8K & 2+0.95K & -0.15K \\
-0.15K & 2+0.95K & 3-1.8K & K+1 \\
b_0 & b_1 & b_2 & \\
b_2 & b_1 & b_0 &
\end{array}
$$

where

$$b_0 = -0.9775K^2 - 2K - 1$$

$$b_1 = 1.6575K^2 - 1.5K - 3$$

$$b_2 = -0.68K^2 - 3.4K - 2$$

Then for stability we need $F(1) > 0$, $F(-1) < 0$, and $|a_0| < a_3$, as well as $|b_0| > |b_2|$. $F(1) = 6 > 0$ and $F(-1) = -3.9K < 0$ implies that $K > 0$, which is what we are assuming anyway. Then $|a_0| < a_3$ implies that $0.15K < |1 + K| = 1+K$ for positive K, or $K > -1.176$, which is satisfied. And $|b_0| > |b_2|$ implies that

$$0.9775K^2 + 2K + 1 > 0.68K^2 + 3.4K + 2$$

$$0.2975K^2 - 1.4K - 1 > 0 \text{ and } K^2 - 4.7K - 3.367 > 0$$

Factoring the left-hand side of this, we get $(K - 5.33)(K + 0.63) > 0$. Therefore, for stability we need $K > 5.33$. In fact at the boundary of the unit circle, $K = 5.33$. With this value of K the closed-loop poles are calculated, using the MATLAB ROOTS(x) statement, to be 0.1263, $0.4580 \pm 0.8889j$. The last two roots have magnitudes of approximately 1.0 and are on the unit circle.

Now let us assume that we want to design the system with the dominant poles having a damping ratio of 0.5. We want to locate the spiral in the z-plane corresponding to $\xi = 0.5$ and then find the value of K where the root locus crosses that line. This can be done graphically using the MATLAB overlay called ZGRID which sets up a graphics window by putting a grid of the unit circle in the z-plane over the root locus. Then use the RLOCFIND statement which allows us to select the desired roots in the z-plane and the value of K is determined automatically at those roots. Instead of just using RLOCUS we type

ZGRID ('new')
RLOCUS (NUM, DEN)
[K, POLES]=RLOCFIND (NUM, DEN)

and MATLAB produces the plot shown in Fig. 4.17 on which we can position the cross-hairs at any point and get the closed-loop pole values and the value of K at that point. For instance, at the point where the root locus crosses the unit circle, we get $K = 5.432$ and a pole at $0.4611+0.8816j$, values very close to those calculated by the Jury method. Locating the $\xi = 0.5$ line as the fifth line up from the real axis in the first quadrant, we position the cross-hairs where that line crosses the root locus. Here MATLAB indicates a value of $K = 45.8436$ and the closed-loop poles at $0.7327 \pm 0.3088j$ and 0.2322.

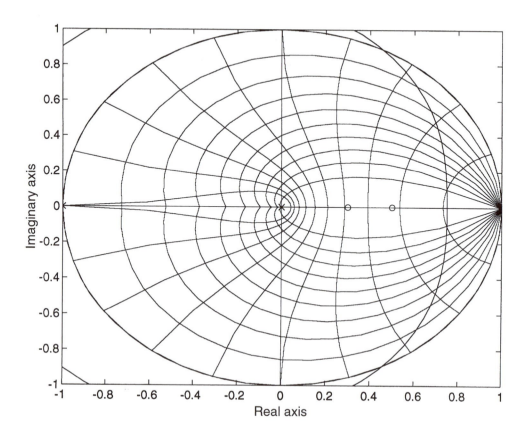

Figure 4.17   ZGRID plot for Example 4.8.

Example 4.9

In Example 4.8, GH (1)=0 and the steady-state error is fixed at 1.0 for unit step inputs. We would often want a smaller value of e_{ss}, say $e_{ss} < 0.1$. To make it possible to accomplish that, assume that the zero in GH (z) is shifted from z = 1 to z = 0.8. Thus

$$GH(z) = \frac{(z-1)(z-0.3)(z-0.8)}{z(z+1)(z+2)}$$

The root locus is plotted in Fig. 4.18.

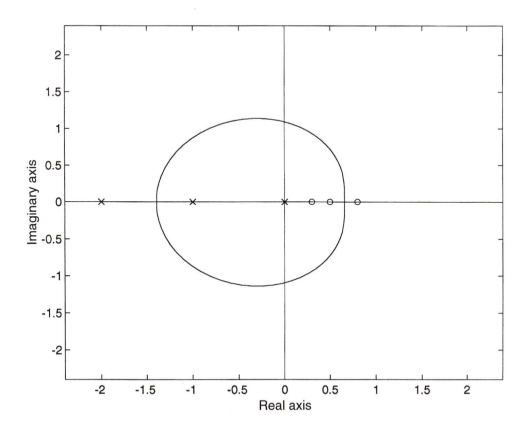

Figure 4.18 Root-locus plot for Example 4.9.

Let us assume that we want a steady-state error less than 0.1:

$$e_{ss} = \frac{1}{1 + KGH(1)} = \frac{1}{1 + 0.0117K} < 0.1$$

from which we get $769.23 < K$. But for such a large value of K we need to locate the closed loop poles on the real axis with $\xi = 1.0$. This can be seen using the ZGRID overlay and the RLOCFIND statement and several trials indicated in Fig. 4.19.

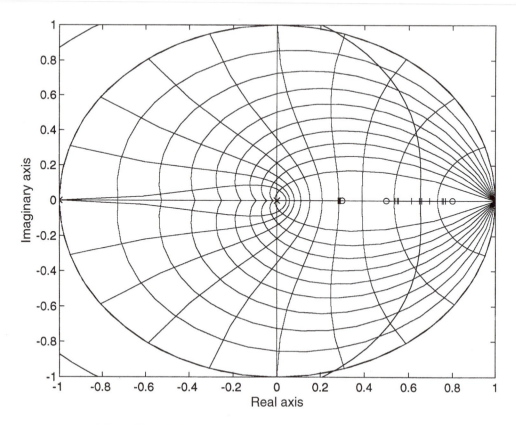

Figure 4.19   ZGRID plot for Example 4.9.

What this example indicates is that it is impossible to meet both the steady-state spec ($e_{ss} < 0.1$) and the transient spec ($\xi = 0.5$) by just varying the gain factor. We need to use compensation techniques. That complicates the problem but also allows much more flexibility. The basic idea of compensation is that in the place of the K parameter we substitute a transfer function and vary the poles and zeros as well as the gain of the transfer function. Before we investigate design via compensators, following a brief consideration of Bode design techniques, we present a final root-locus example.

Example 4.10

Here we consider a more complex root-locus. A hand plot would be tedious. But a MAT-LAB construction is painless. Let KGH(z) take the form

$$KGH(z) = K\frac{z(z-1)(z+1)(z-2)(z+2)}{(z-0.5)(z+0.2)(z+0.3)(z-0.3)(z-0.8)}$$

A MATLAB root locus plot appears in Fig. 4.20. The ZGRID overlay is presented in Fig. 4.21 and we see that as the poles go to the zeros, the root locus indicates that beyond certain values of K the system goes unstable.

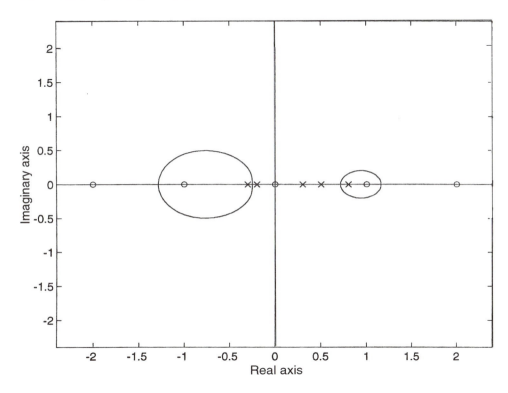

Figure 4.20   MATLAB root-locus plot for Example 4.10.

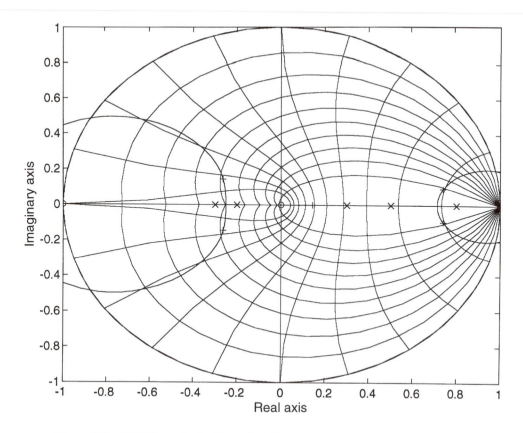

Figure 4.21   ZGRID overlay for Example 4.10.

With the closed-loop system having five poles, where to place them via K selection becomes an extremely difficult task. For instance, if we select the first and fourth quadrant poles to have a 0.9 damping ratio (as indicated by the cross-hairs in Fig. 4.21), there are two other poles in the second and third quadrants with a damping ratio of about 0.4, which indicated a much more underdamped response. It will matter which set of complex poles dominate. The ZGRID constant ξ and constant ω_n contours apply only, strictly speaking, to second-order systems. So from the root-locus plot alone it is very difficult to design these higher-order systems. Time responses would have to be looked at in conjunction with K value root-locus selections. This points to the need for compensation techniques as well as the more sophisticated pole placement design approaches that are taken up in Chapter 6.

Bode Plots

Bode plots are graphs of the magnitude and phase angle of a system's open-loop transfer function. We saw in Chapter 2 that applying a bilinear transformation

$$z = \frac{1 + \dfrac{T}{2}w}{1 - \dfrac{T}{2}w} \qquad (4.32)$$

to the z variable in a transfer function G(z) generated a function G(w) which could be treated with the familiar Bode construction techniques employed for continuous systems. This is because the unit circle of the z-plane transforms under this bilinear transformation into the imaginary axis of the w-plane. Assume that we have transformed a given G(z) to G(w). Along the unit circle in the z-plane $z = e^{j\omega T}$ and

$$w = j\frac{2}{T}\tan(\frac{\omega T}{2}) = j\omega_w \qquad (4.33)$$

We are actually interested in the gain and phase plots of G ($j\omega_w$). We refer to ω_w as the w-plane frequency. Since $\omega_w = j\left(\frac{2}{T}\right)\tan\left(\frac{\omega T}{2}\right)$, if $\frac{\omega T}{2}$ is small, then $\omega_w \approx \left(\frac{2}{T}\right)\left(\frac{\omega T}{2}\right) = \omega$. The higher we choose the sampling frequency, of course, the better this approximation will be. If it can be afforded, 20 times the highest frequency in the system is an excellent choice. Although CAE tools are available to generate Bode graphs of a given G(z) directly – and will be looked at shortly – it is useful to be able quickly to sketch Bode plots of low order systems using familiar continuous system techniques. The following steps review the procedure for constructing Bode plots of a system's open loop transfer function G($j\omega_w$).

Assume that G($j\omega_w$)=$|G(\omega_w)|$ arg G(ω_w) = $|G(\omega_w)|\phi(\omega_w)$ and consider the following types of terms:

1. Constant gain K (Fig 4.22)
   - Magnitude = $20\log_{10}K$
   - Slope = 0
   - Phase angle $\phi(\omega_w) = 0$

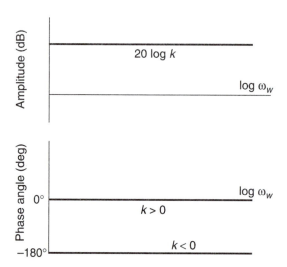

Figure 4.22   Bode plots for a constant gain K.

2.   Pole at the origin $j\omega_w$ (Fig. 4.23)

- Magnitude = $20\log_{10}\left|\dfrac{1}{j\omega_w}\right| = -20\log_{10}\omega_w$

- Slope = -20 dB/decade

- Phase angle = $\phi(\omega_w) = -90°$

or, in general, for multiple poles at the origin:

- Magnitude = $20\log_{10}\left|\dfrac{1}{(j\omega_w)^n}\right| = -20n\log_{10}\omega_w$

- Slope = -20n dB/decade

- Phase angle = $\phi(\omega_w) = -90°n$

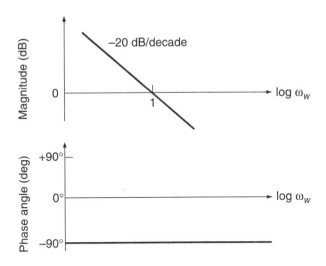

Figure 4.23 Bode plots for a pole at the origin.

3. A zero at the origin (Fig. 4.24)

- Magnitude = $20\log_{10}\left|j\omega_w\right|$

- Slope = $20\,{}^{db}\!\!\big/\!_{decade}$

- Phase angle = $\phi(\omega_w) = +90^\circ$

or, in general, for multiple zeros at the origin:

- Magnitude = $20\log_{10}\left|(j\omega_w)^n\right| = 20n\log_{10}\omega_w$

- Slope = $20n\,{}^{db}\!\!\big/\!_{decade}$

- Phase angle = $\phi(\omega_w) = 90^\circ n$

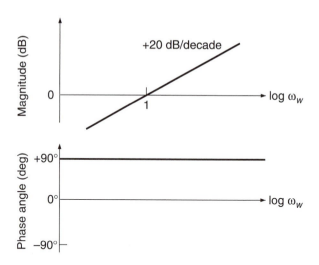

Figure 4.24   Bode plots for a zero at the origin.

4.   Pole on the real axis (Fig. 4.25)

- Magnitude = $20\log_{10}\left|\dfrac{1}{1+j\omega_w\tau}\right| = -10\log_{10}\left(1+\omega_w^2\tau^2\right)$

- Slope = $-20\,{}^{db}\!\!\diagup\!_{decade}$

- Phase angle = $\phi(\omega_w) = -\tan^{-1}\omega_w\tau$

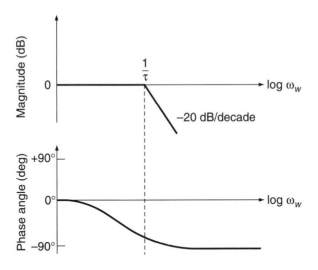

Figure 4.25 Bode plots for a pole on the real axis.

5. Zero on the real axis (Fig. 4.26)

- Magnitude = $20 \log_{10} |1 + j\omega_w \tau| = 10 \log_{10} (1 + \omega_w^2 \tau^2)$

- Slope = $20 \, \mathrm{db}/\mathrm{decade}$

- Phase angle = $\phi(\omega_w) = \tan^{-1} \omega_w \tau$

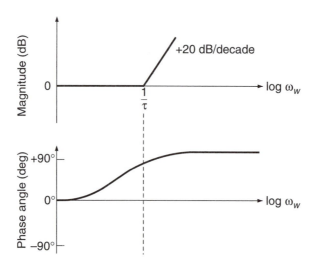

Figure 4.26   Bode plots for a zero on the real axis.

Example 4.11

For the open-loop transfer function

$$G(z) = \frac{0.5z + 0.2}{z^2 - 1.4z + 0.4}$$

we can rewrite $G(z)$ as $G(w)$, letting $T = 2$ so that $z = \dfrac{(1+w)}{(1-w)}$ and

$$G(w) = G(z)\Big|_{z=\frac{1+w}{1-w}} = \frac{0.5\dfrac{(1+w)}{(1-w)} + 0.2}{\dfrac{(1+w)^2}{(1-w)^2} - 1.4\dfrac{(1+w)}{(1-w)} + 0.4}.$$

A little algebra reduces this to

$$G(w) = \frac{-0.3w^2 - 0.4w + 0.7}{1.2w} = \frac{-0.3(w + 2.33)(w - 1)}{1.2w}$$

By applying the procedures above to G(w), we can obtain the magnitude and phase angle plots. It is even easier to use the BODE statement from MATLAB, which yields Bode plots for analog transfer functions, as shown in Fig. 4.27.

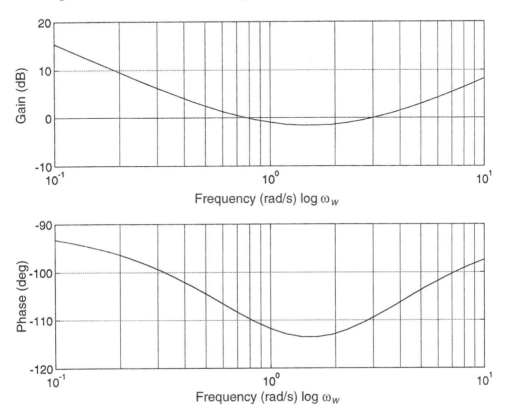

Figure 4.27   Magnitude and phase angle plots of Example 4.11.

In Section 4.1 we defined control system design parameters related to steady state and transient behaviors. Here we define two important frequency domain design parameters, *phase margin* and *gain margin,* which are relative stability indicators.

Definition 4.1. The phase margin of a discrete system determines the amount of additional phase that can be added to the system before it goes unstable:

$$\text{phase margin} = \phi_m = [180^\circ + \arg GH(e^{j\omega_1 T})] \tag{4.34}$$

where $\left| GH(e^{j\omega_1 T}) \right| = 1$ and ω_1 = gain crossover frequency. Phase margin for a second-order system is related to the damping ratio ξ by

$$\phi_m = \tan^{-1}\left[\frac{2\xi}{\left(\sqrt{4\xi^4 + 1} - 2\xi^2\right)^{\frac{1}{2}}} \right] \tag{4.35}$$

which is approximately $\phi_m \cong 100\xi$.

Definition 4.2. The gain margin of a system determines the amount of extra gain the system tolerates before it goes unstable. The gain margin (GM) is defined as the magnitude in decibels of the reciprocal of the open-loop transfer function GH ($e^{j\omega T}$), evaluated at the frequency ω_π, which is the frequency at which the phase angle of GH ($e^{j\omega T}$) is $-180°$. ω_π is called the phase crossover frequency.

$$\text{GM} = \text{gain margin} = 20\log \frac{1}{\left| GH(e^{j\omega_\pi T}) \right|} \tag{4.36}$$

where $\arg GH(e^{j\omega_\pi T}) = -180°$.

To locate the GM and ϕ_m we need to find the frequency ω_1 where the gain is 0 dB and the frequency ω_π where the phase is $-180°$. These are usually easy to locate on Bode plots. We read the phase at ω_1 and the gain at ω_π and these give us, respectively, the phase margin and the gain margin as indicated in the sketches in Fig. 4.28.

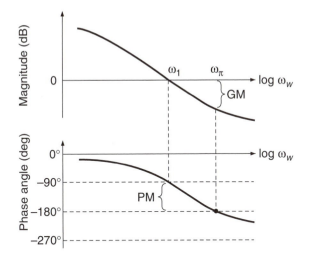

Figure 4.28 Typical phase and gain margins.

Although it is useful to be able to plot G(z) Bode plots as G(w) plots using familiar analog techniques, the recent proliferation of CAE tools capable of plotting G(z) directly is making the G(w) approach to discrete system design less and less attractive. The MATLAB statement DBODE, for instance, will calculate and plot Bode plots for any given G(z).

Example 4.12

Use MATLAB to generate the Bode plots for Example 4.4.

Solution:

$$G(z) = \frac{0.5z + 0.2}{z^2 - 1.4z + 0.4}$$

and all we need to do is write

```
T=2
NUM=[0.5 0.2]
DEN=[1 -1.4 0.4]
DBODE (NUM, DEN, T)
```

and MATLAB returns the plot in Fig. 4.29.

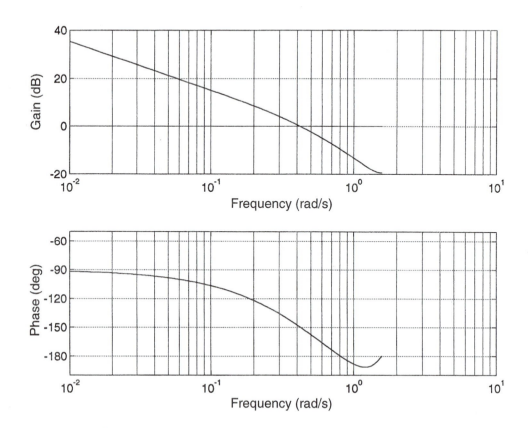

Figure 4.29   Gain and phase plots for Example 4.12.

If we compare this plot to the corresponding G(w) plot in Fig. 4.27, we note that they are similar for low frequencies but deviate considerably at higher frequencies. This is because of the relatively large value of T. If T is decreased from T = 2 seconds to T = 0.2 second, the G(w) and G(z) Bode plots look quite similar (Fig. 4.30 and 4.31).

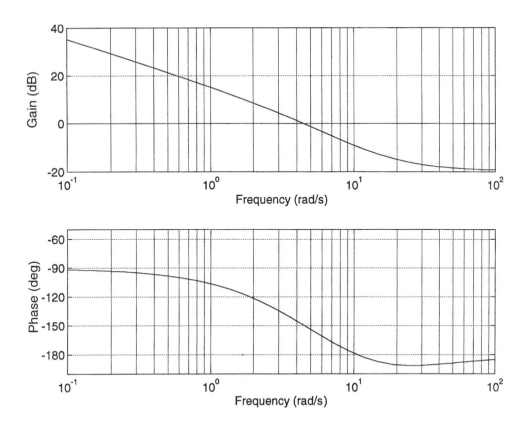

Figure 4.30   Bode plot of G(w) for T = 0.2.

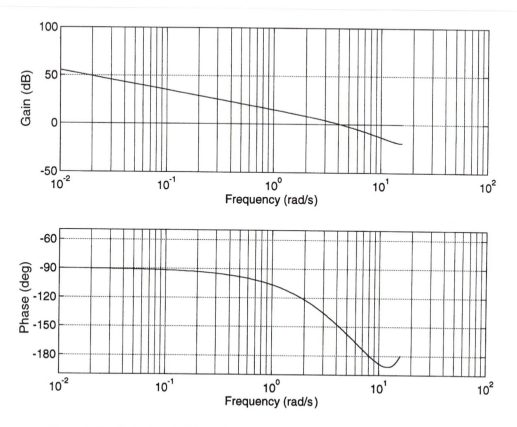

Figure 4.31   Bode plot of G(z) for T = 0.2.

For both of these plots the GM and ϕ_m values are quite similar. Approximately, GM = 10 dB and ϕ_m = 30 degrees. However, for the G(w) plot in Fig. 4.27 we get approximately, GM =∞ dB and ϕ_m = 70 degrees and for the G(z) plot of Fig. 4.29 GM = 10 dB and ϕ_m =30 degrees. Care must of course be taken when using w-plane transformations, especially if the sampling rates are low.

For Bode design problems in general, KGH is plotted either by hand as KGH(w) or by machine as KGH(z) assuming that K = 1. Then if K is increased or decreased, the effect is to raise or lower the entire magnitude plot by 20 log KdB. These vertical shifts in the magnitude plot will then indicate lower or higher gain and phase margins.

Example 4.13

Use GH (z) = G (z) from Example 4.12 with T = 0.2 second and determine the value of K necessary to have 20 dB of gain margin.

Solution: From the plot in Fig. 4.31 we have approximately 10 dB of gain margin implying that we would like the gain curve lowered by 10 dB so that at $\omega_\pi = 0.8 \, \text{rad} / \sec$ we have a gain of –20 dB. That implies that $20 \log K = -10$ or $K = 0.32$ instead of the nominally assumed value of $K = 1$. In this case the phase margin is also increased to 60 degrees. The problem may be that such a low value of K means that the system does not meet other specs.

Example 4.14

Let us design the system of Example 4.13 so that it has a damping ratio $\xi = 0.75$ and also a steady-state error less than 0.1 for ramp inputs.

Solution: From Equation 4.35 the phase margin should be about 75 degrees. From Fig. 4.31 that implies we should have the gain crossover frequency $\omega_1 \approx 10^0 = 1.0 \, \text{rad} / \sec$. Then we need to lower the gain curve approximately 15 dB. Therefore, $20 \log K = -15$ or $K = 0.18$.

Now look at e_{ss}. Our system is type 1 since it has a pole at z=1. From Table 4.1, $e_{ss} = \dfrac{1}{K_v}$ where

$$K_v = \frac{1}{T} \lim_{z \to 1}(z-1)KGH(z) = \frac{1}{0.2} K \frac{0.5+0.2}{1-0.4} = 5.83K$$

Therefore, to have $e_{ss} < 0.1$ we need $K > 1.71$, but to meet the transient spec on the damping ratio we had to lower K from 1.0 to 0.18. So both specs cannot be met simultaneously with just K variation. Compensation is called for.

4.4 COMPENSATION

A component $G_c(z)$ that changes the system performance according to a set of desired specifications is called a *compensator* or *controller*. A compensator can be placed in any of several different places in a control system. Fig. 4.32 shows a compensator $G_c(z)$ placed in a complex control system configuration.

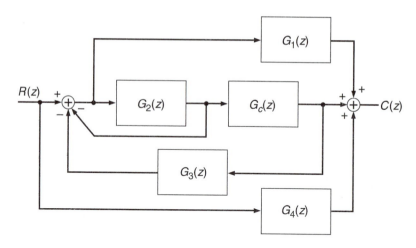

Figure 4.32   Complex compensator configuration.

However, most systems can be reduced or restructured to a more accessible form. Fig. 4.33 shows a compensator $G_c(z)$ used in a basic discrete feedback control system configuration, called the *cascade compensation configuration,* to which most systems can be reduced and which will be our primary focus.

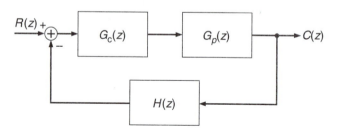

Figure 4.33   Compensator $G_c(z)$ used in a discrete feedback control system.

Although other basic schemes exist, such as feedforward compensation or minor loop compensation, cascade compensation is by far the most common and popular compensation scheme. The overall closed-loop transfer function of this system is

$$T(z) = \frac{G_c(z)G_p(z)H(z)}{1+G_c(z)G_p(z)H(z)} \tag{4.37}$$

An obvious kind of compensator design procedure, called direct design, is simply to specify a $T(z)$ that will yield a desired transient response to an input such as the unit step function. Then if $G_p(z)$ and $H(z)$ are known, we can represent the compensator transfer function as

$$G_c(z) = \frac{T(z)}{G_p(z)H(z)[1-T(z)]} \tag{4.38}$$

and for unity gain feedback,

$$G_c(z) = \frac{T(z)}{G_p(z)[1-T(z)]} \tag{4.39}$$

Example 4.15

Let us assume that we have a plant transfer function of the form

$$G(z) = \frac{2(z+0.5)}{z^2 - z + 0.5}$$

and we want to design a compensator that will yield a closed-loop transfer function of the form

$$T(z) = \frac{z + 0.5}{z^2 + 0.5z + 0.2}$$

From Equation 4.39 we can write

$$G_c(z) = \frac{T(z)}{G(z)[1-T(z)]} = \frac{z^2 - z + 0.5}{2(z^2 - 0.5z - 0.3)}$$

That's all there is to it. We only need to implement the controller, for example, using a DSP.

Although such a compensator design procedure seems to be straightforward, there are several problems associated with it. In particular, we must make sure that the transfer functions $T(z)$ and $G_c(z)$ are both stable and physically realizable. To be physically realizable, a transfer function must not have more zeros than poles. The transfer functions of Example 4.15 were both stable and physically realizable, but often we will specify a stable and physically realizable T(z) and the design yields an unstable $G_c(z)$. Also, pole/zero cancellation can be problematic. For instance, if the plant $G_p(z)$ has any zeros outside the unit circle, they must be canceled by zeros of T(z) or else they will appear as poles of $G_c(z)$ indicating an unstable controller. Several constraint equations can be developed to guarantee a successful design, but the procedure gets cumbersome, and we will not pursue it further. In addition, the pole placement procedure – considered with the state-variable design methods of Chapter 6 – will do essentially the same thing as this design method.

In this section we consider the use of Bode plots to study phase-lead and phase-lag compensators. We discuss a special type of lead-lag compensator called a PID controller.

Phase-Lead and Phase-Lag Compensators

First we review the basic compensation techniques in the continuous domain. A first-order compensator has the transfer function

$$G_c(s) = \frac{k(s+z)}{(s+p)} \tag{4.40}$$

where k is the constant gain and z and p are the zero and pole of the compensator. The design of this type of compensator consists of calculating the gain k and the pole-zero locations on the s-plane. Fig. 4.34 shows the pole-zero pattern of phase-lag and phase-lead compensators in the s-plane.

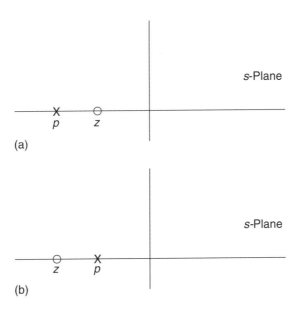

Figure 4.34   Pole-zero patterns of phase-lag and phase-lead compensators: (a) phase lead, $|z| < |p|$; (b) phase lag, $|p| < |z|$.

In a phase-lead compensator, the phase of the output leads the phase of the input. Phase-lead compensators provide a faster response time and generally bring a stabilizing effect to a system, although since high-frequency effects are augmented, the system's closed-loop bandwidth tends to be increased and high-frequency noise can sometimes become problematic. Fig. 4.35 shows the frequency response magnitude and phase angle characteristics of a phase-lead device.

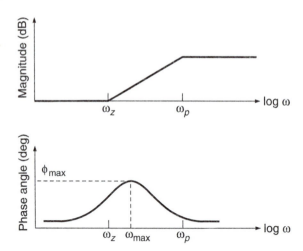

Figure 4.35   Magnitude and phase angle of a phase-lead device.

Fig. 4.36 shows the magnitude and phase angle of a phase-lag device. In a phase-lag compensator, the phase of the output lags the phase of the input, which means a negative phase bulge is contributed by G_c when G_c is multiplied by G_p to form the

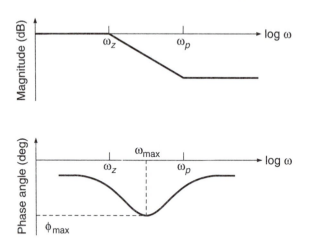

Figure 4.36   Magnitude and phase angle of a phase-lag device.

open-loop transfer function. Care must be taken that the peak of the phase bulge is well below (about a decade of frequency is usually good) the gain crossover frequency. Phase-lag compensators tend to provide higher loop gain and attenuation but reduce the closed-loop system's bandwidth.

   The general design procedure in the z-domain for phase-lead and phase-lag compensators resembles the one in the s-domain. Corresponding to the continuous s-domain compensator transfer function $G_c(s)$, indicated by Equation 4.40, we can write

$$G_c(z) = \frac{K(z - z_1)}{(z - z_2)} \tag{4.41}$$

In some cases we simplify our design if the dc gain can be fixed. If, for example, we let the dc gain be unity, then $G_c(1) = 1$ and

$$K = \frac{1 - z_2}{1 - z_1} \tag{4.42}$$

and we have to determine only two parameters, z_1 and z_2, to complete the design. The dc gain may need to be shifted from unity, but a value of unity is often a good point of departure for a first cut at the design.

   If we translate the compensator transfer function into the w-plane, we get

$$G_c(w) = \frac{\alpha(1 + w / \omega_z)}{(1 + w / \omega_p)} \tag{4.43}$$

where ω_z and ω_p are, respectively, the zero and pole of the compensator transfer function. The α value is the low-frequency gain in the ω_w space. Once we have made the translation, w-plane design procedures, mimicking classical analog design, can profitably be employed in the design of discrete systems. The w-plane parameters relate to the z-plane parameters as follows:

$$
\begin{aligned}
z_1 &= \frac{\dfrac{2}{T} - \omega_z}{\dfrac{2}{T} + \omega_z} \\[2em]
z_2 &= \frac{\dfrac{2}{T} - \omega_p}{\dfrac{2}{T} + \omega_p}
\end{aligned}
\tag{4.44}
$$

$$K = \alpha \left[\frac{\omega_p (\omega_z + \frac{2}{T})}{\omega_z (\omega_p + \frac{2}{T})} \right]$$

There are several popular z- and w-plane design procedures, all requiring greater or lesser degrees of trial and error. As an illustration we present the following procedure for lead compensation.

Lead compensation procedure

1. Using the MATLAB expression DBODE, plot the frequency response of the given $G_p(z)$ transfer function.

2. Evaluate the uncompensated system gain crossover frequency (ω_1) and phase crossover frequency (ω_π) and gain and phase margins from the DBODE plot. Note that the DBODE plots provide frequency response along the ω axis, whereas the compensator frequency response is plotted along the ω_w axis. We assume, though, that both frequencies are roughly equivalent.

3. To reshape the frequency response we want to place the phase bulge maximum (ϕ_{max}) of the phase-lead device at a frequency that would be the new gain crossover frequency (ω_{1x}) of the compensated system. The problem is that we have no simple way to determine ω_{1x}. Our experience indicates a good initial guess is to let $\omega_{1x} = 1.5\omega_1$. Then we may need to iterate to get the desired response.

4. How much phase do we want to add into the system? A good initial strategy is to assume that we want ϕ_{max} to be 37 degrees because from the equation $\tan \phi_{max} = \frac{1}{2}\left\{ \mu - \left(\frac{1}{\mu} \right) \right\}$ we then get that $\mu = 2$ which means that we place the break frequencies of the phase-lead compensator at $\omega_z = \left(\frac{1}{\mu} \right)\omega_{1x}$ and $\omega_p = \mu\omega_{1x}$, an octave above and an octave below the ω_{1x} value.

5. If the resulting design has insufficient phase margin, we can increase μ, which will spread out the break frequencies and increase the value of ϕ_{max}.

6. Once we decide on values for the break frequencies, we use Equation 4.44 to solve for z_1 and z_2 and the gain K assuming initially that $\alpha = 1$. Or we may wish to use Equation 4.42 for an initial guess of the K value. Of course, we

must also specify the sampling period T. Then construct the compensator transfer function

$$G_c(z) = \frac{K(z - z_1)}{(z - z_2)}$$

from Equation 4.41.

7. Check gain and phase margins by using DBODE to plot the frequency response of $G_c(z)G_p(z)$. Iterations will generally be necessary, especially on α or K or ϕ_{max}.

Example 4.16

Consider the plant

$$G_p(z) = \frac{0.5(z + 0.2)}{z(z - 0.5)^2}$$

with sampling period of 0.05 second. Let us assume that we want this system to be compensated so that the closed-loop system has a phase margin of about 60 degrees. Using DBODE we construct the frequency response plot in Fig. 4.37.

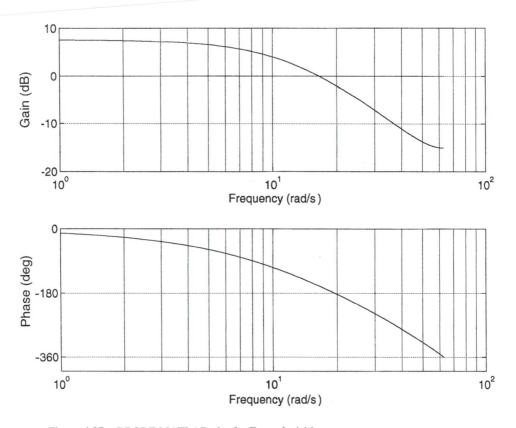

Figure 4.37   DBODE MATLAB plot for Example 4.16.

From this plot we can read the gain and phase margins as approximately $\phi_m = 30°$ and GM = 2.5 dB. The gain crossover frequency is $\omega_1 \doteq 17 \text{ rad/sec}$ so we choose as an initial guess $\omega_{1_x} = 25.5 \text{ rad/sec}$.

Try $\omega_z = \dfrac{1}{\mu}\,\omega_{1x}$ and $\omega_p = \mu\,\omega_{1x}$ with $\mu = 2$, or $\omega_z = 12.75$ and $\omega_p = 51$ rad/sec. Then using Equation 4.44 with $\alpha = 1$ we get

$$z_1 = 0.531$$
$$z_2 = -0.121$$
$$k = 2.32$$

and the compensator transfer function becomes

$$G_c(z) = 2.32 \frac{(z - 0.531)}{(z + 0.121)}$$

If we multiply this by $G_p(z)$ and construct the Bode plot using DBODE, we get the plot in Fig. 4.38. From this plot we note that the low-frequency gain is maintained from the uncompensated system but the system is unstable. The first thing to try is cutting the gain in half. That yields the plot in Fig. 4.39. Here we note a considerable improvement. The gain margin is about 4 dB and the phase margin 100 degrees but the low-frequency gain is attenuated to about 2 dB. Increasing the gain slightly to 0.7 (which is 0.5K), we get the plot in Fig. 4.40.

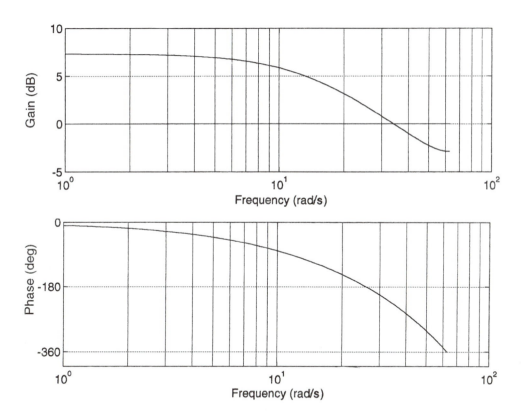

Figure 4.38   DBODE MATLAB plot for $G_p(z)\,G_c(z)$.

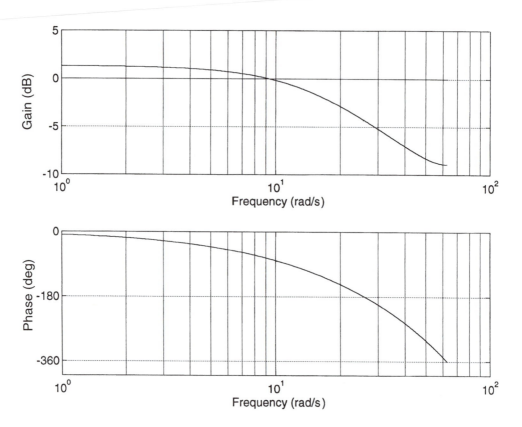

Figure 4.39   Plot of $G_p(z)G_c(z)$ with gain halved.

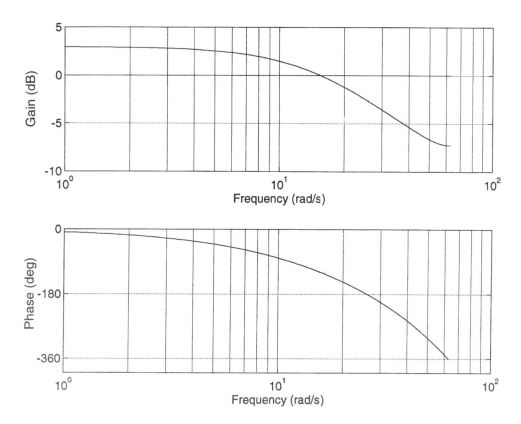

Figure 4.40 Plot of $G_p(z)G_c(z)$ with gain = 0.7.

Here we note a phase margin of 65 degrees, which meets the design specification, and a gain margin of about 3 dB, slightly better than the 2.5 dB gain margin of the uncompensated system. Also, the low-frequency gain is not attenuated as much as in the preceding case.

Again, this type of design, like all classical frequency-domain procedures, entails a lot of trial and error. The more modern approaches to design generally require less trial and error, but at the expense of more complicated mathematics. Modern design employing state-space representations is taken up in Chapter 6.

In addition, in the lead compensation discussed here, and also in lag compensation schemes, the amount of massaging of the open-loop transfer function that is possible is quite limited. Other compensation schemes allow more aspects of the frequency response to be manipulated. Because both lag and lead compensators each have advantages and disadvantages, the two are cascaded together in some designs. The resulting compensator is called a *lead-lag compensator*. The lag part provides low frequency gain, and the lead part brings stability margins to the system. We consider as a special case of the general lead-lag compensator the proportional-integral-derivative (PID) controller.

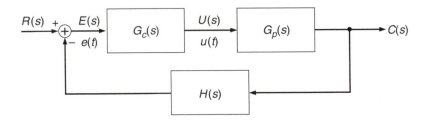

Figure 4.41   Control system with a compensator.

PID Compensator

As in the system shown in Fig. 4.41, we can represent the compensator $G_c(s)$ as a transfer function with the error signal E(s) as input and U(s) as output. We consider continuous-time systems to develop the PID control scheme, then translate into the discrete domain. If

$$G_c(s) = K_p \qquad (4.45)$$

or

$$u(t) = K_p \, e(t) \qquad (4.46)$$

we have a proportional controller. Here the controller can introduce a large gain and therefore decrease the rise time. As a result, the output follows the input faster. The problem with a large gain in a proportional controller is increased overshoot. The increased overshoot can produce signals that drive the system into saturation, bringing into play nonlinear effects that degrade system performance. Proportional control alone is usually insufficient.

A derivative control is used to reduce the rise time without increasing percent overshoot, so this type of controller improves system stability. The transfer function of this controller is the differential of the error term

$$u(t) = K_D \frac{de(t)}{dt} \qquad (4.47)$$

or

$$G_c(s) = sK_D \qquad (4.48)$$

The use of a controller of the form of Equation 4.48 is rare. However, if this and the previous controller are combined, we get a proportional-derivative (PD) controller, which is a popular type of compensation scheme. One major problem with a PD controller is the large steady-state errors that may result.

To reduce the steady-state error, we use an integral control. Then

$$u(t) = K_I \int_0^t e(\tau)d\tau \qquad (4.49)$$

or

$$G_c(s) = \frac{K_I}{s} \qquad (4.50)$$

Although an integral control is occasionally used by itself, a more common scheme is to combine the integral with a proportional term to get the proportional-integral (PI) controller.

More generally, and in many control system applications, such as servomotor control systems, the combination of proportional, integral, and derivative control are used in the famous PID controller. A PID controller has the transfer function

$$\frac{U(s)}{E(s)} = G_c(s) = K_P + K_D s + \frac{K_I}{s} \qquad (4.51)$$

where K_P, K_D, and K_I are the gain constants of the controller.

By using the bilinear transformation, we can transform the transfer function of each controller from the continuous domain to the discrete domain. Table 4.2 shows the transfer function of a proportional integral (PI) controller, a proportional derivative (PD) controller, and a PID controller in the z-domain. In this table

Table 4.2   Transfer Functions of PI, PD, and PID Controllers

Controller	PD	PI	PID
Transfer Function	$\dfrac{A_1 z + A_2}{z}$	$\dfrac{B_1 z + B_2}{z - 1}$	$\dfrac{C_1 z^2 + C_2 z + C_3}{z^2 - z}$

$$A_1 = K_P + \frac{K_D}{T} \text{ and } A_2 = \frac{-K_D}{T} \qquad (4.52)$$

$$B_1 = K_P + K_I T \text{ and } B_2 = -K_P \qquad (4.53)$$

and

$$C_1 = K_P + K_I T + \frac{K_D}{T}, \; C_2 = -(K_P + 2\frac{K_D}{T}), \; C_3 = \frac{K_D}{T} \qquad (4.54)$$

    Designing a PID controller in the z-domain requires determining the gain constants C_1, C_2, and C_3 and these in turn require K_P, K_I, K_D determinations and an assumed value for T. As with lead and lag compensators, there are several useful design approaches to PID controllers. We mention only one, an empirical procedure based on the Ziegler-Nichols method. The Ziegler-Nichols method of tuning a PID compensator is used widely in industry, especially for large-scale or complex systems, because it does not require having an accurate plant model.

PID design procedure

1. We initiate our procedure by multiplying the given $G_p(z)$ by the PID transfer function

$$G_c(z) = K_P + K_D \frac{(z-1)}{Tz} + K_I \frac{Tz}{(z-1)} \qquad (4.55)$$

    Set $K_D = K_I = 0$, then gradually increase K_P until the system goes unstable (gets poles on the unit circle in the z-plane).

2. At this value $K_P = K_x$ and we note the frequency at which the system goes from being stable to being unstable, the frequency of oscillation, ω_x.

3. Then, in the PID compensator, we set

$$K_P = 0.6K_x$$

$$K_I = \frac{K_P \omega_x}{\pi} \qquad (4.56)$$

$$K_D = \frac{K_P \pi}{4\omega_x}$$

4. Fine tuning can be achieved by iterating on the K_I and K_D values, gradually decreasing K_I and increasing K_D.

Example 4.17

Consider the system from Example 4.16 with plant

$$G_p(z) = \frac{0.5(z+0.2)}{z(z-0.5)^2}$$

with sampling period of 0.05 second. Design a PID controller for this system that meets the 60-degree phase margin achieved with a lead controller.

Solution: First, we need to assume that $G_c(z) = K_P$ and consider increasing K_P and note the effect in the Bode plot of $G_c(z)G_p(z)$. (If the plant model were not available, we could drive the system with a step input and gradually increase the K_P gain until we noted oscillation at the output.) From the Bode plot we note where the gain margin and phase margins shrink to zero and the gain and phase crossover frequencies become equal. That frequency is ω_x. The Bode plot with $K_P=1$ is the plot of the uncompensated system shown in Fig. 4.37. We noted there a gain margin of about 2.5 dB and a phase margin of 30 degrees. Increasing K_P and looking at several plots, we arrive at the DBODE plot in Fig. 4.42 and note that here $K_x \approx 1.24$ and we read from the plot $\omega_x \approx 20 \text{ rad / sec}$.

Then using Equation 4.56, we calculate

$$K_P = 0.744$$
$$K_I = 4.736$$
$$K_D = 0.029$$

We want to check the closed-loop system unit step response, which can be done using the DSTEP statement from MATLAB. Since we envision several trial-and-error attempts, the m-file approach from MATLAB is employed. The m-file allows us to easily do repeated

runs on a problem that might otherwise be computationally tedious. We use Equation 4.54
to calculate the numerator coefficients in the PID controller and write

```
pid.m
c1=kp+ki*t+kd/t
c2=-(kp+2*kd/t)
c3=kd/t
N=[0 0 0.5*c1 0.5*c2+0.1*c1 0.1*c2+0.5*c3 0.1*c3]
D=[1 -2 1.25 -0.25 0 0]
DSTEP (N, N+D)
end
```

The N and D terms are the products of the numerators and denominators respectively
of the plant and controller transfer functions. We enter the m-file with $T = 0.5$ and the K
values indicated above and get the step response indicated in Fig. 4.43. We note quite a bit
of overshoot in this case, and, in accordance with the design procedure, a decrease in K_I is
suggested. Trying $K_I = 4.0$ yields the plot in Fig. 4.44, indicating less overshoot. A
value of 3.0 was even better, and a value of 2.0, indicated in Fig. 4.45, was better still.

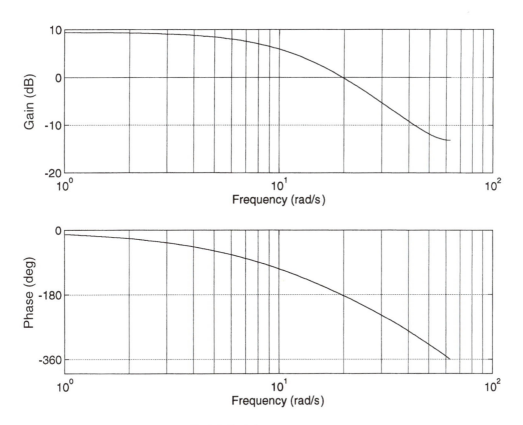

Figure 4.42 Bode plot of $G_p(z)G_c(z)$ indicating oscillation.

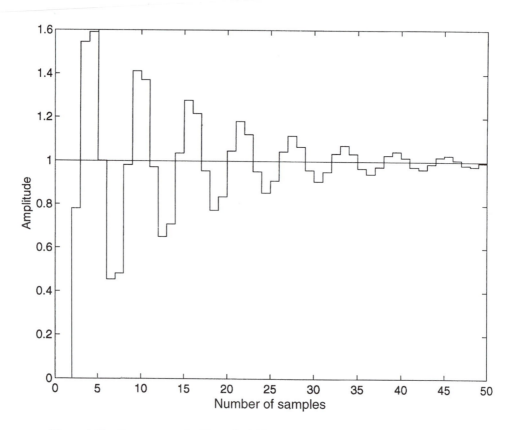

Figure 4.43   Step response for Example 4.17.

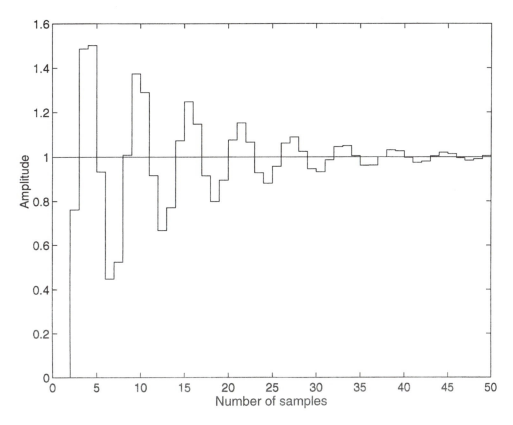

Figure 4.44   Step response with $K_1 = 4.0$.

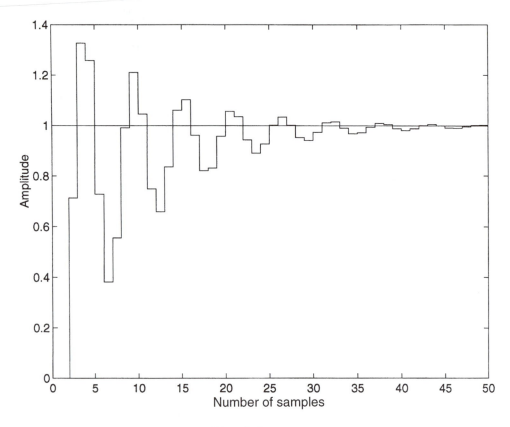

Figure 4.45   Step response with $K_I = 2.0$.

Further reductions in K_I were not helpful. More overshoot appeared and the time to settle out was increased. Then we tried to slightly increase the K_D value but got more overshoot, so we decided that $K_I = 2.0$, $K_D = 0.029$, and $K_P = 0.744$ looked pretty good. In fact, using these values and plotting the open-loop frequency response, we get the plot in Fig. 4.46, which is a blow-up over the frequency range of interest.

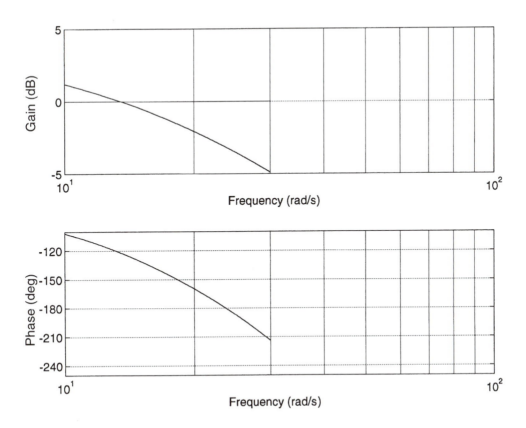

Figure 4.46   Bode plot for PID compensator.

We note a gain margin of about 2.5 dB and a phase margin of about 55 degrees which are close enough to our desired values. One other point of interest regarding PID design: Note that the steady-state error in response to the step inputs is zero, unlike in the lead compensator design, where even though we get similar gain and phase margins, we must live with a nonzero steady-state error.

Once the gains C_1, C_2, C_3 are decided on, we can determine the difference equation of the PID controller from the transfer function. We can then create a PID algorithm and implement it by programming assembly language subroutines for a DSP chip. In Section 5.4 we discuss some of the implementation considerations in the design of PID controllers.

Deadbeat Controllers

One more type of digital compensator will be looked at in this section. A special type of controller, called a *deadbeat controller,* responds quickly to the control input and settles out with zero error in minimum finite time. A deadbeat controller can be designed in the z-domain by replacing the closed-loop poles of the system with poles at the origin of the z-domain. This type of controller design is actually a special case of the direct design with which we began this section on compensation. Deadbeat controllers, then, suffer from all the problems and constraints that need consideration when doing direct design. In addition, the controllers resulting from this design procedure tend to be highly sensitive to parameter variations within the system. However, deadbeat design presents an optimal kind of design (optimal in the sense of minimal settling time) and is often useful for comparison purposes.

We consider only the most elementary formulation. Assume that the plant transfer function has all poles and zeros inside the unit circle. Then we can use Equation 4.39 repeated here

$$G_c(z) = \frac{T(z)}{G_p(z)[1 - T(z)]} \tag{4.57}$$

We want T(z), the closed-loop transfer function, to have all its poles at zero and assuming a unit step input $r(n) \leftrightarrow R(z) = \dfrac{z}{(z-1)}$, we can let $T(z) = \dfrac{1}{z^k}$ if the number of poles of $G_p(z)$ exceeds the number of zeros by k.

Example 4.18

A discrete-time system has a plant transfer function

$$G_p(z) = \frac{0.5z^{-1} + 0.6z^{-2}}{1 - 0.2z^{-1} + 0.2z^{-2}}$$

We can design a deadbeat controller for this transfer function using $T(z) = \dfrac{1}{z^k}$ with k = 1 substituted in Equation 4.57. Solving for $G_c(z)$ we get

$$G_c(z) = \frac{z^2 - 0.2z + 0.2}{(z-1)(0.5z + 0.6)}$$

which is physically realizable. What a transfer function of $T(z) = \dfrac{1}{z}$ implies is that

$$T(z) = \frac{C(z)}{R(z)} = \frac{1}{z}$$

and

$$C(z) = R(z)/z = \frac{z}{z-1}\frac{1}{z} = \frac{1}{z-1} = z^{-1} + z^{-2} + \cdots$$

which translates into the time domain as $c(n) = u(n-1)$ or the output is equal to the input shifted by one sampling period.

This ability of a discrete system to reach steady state in a finite number of sampling periods is unique to discrete systems. A linear continuous system, even if it is stable, requires infinite time to reach steady state asymptotically. However, with the design procedure outlined here, we cannot be certain that the deadbeat system will be free of ripples in between samples. When the discrete signals are converted to their continuous representations in a sampled data system, we may experience some unanticipated ripples. More complicated design procedures are called for to guarantee ripple-free deadbeat designs. The modified z-transform may be employed. Or the procedure presented in Kuo [12] based on the work of Sirisena is particularly appealing because of its mathematical elegance.

4.5 SUMMARY

In this chapter we introduced tools and techniques for designing digital control systems. Design parameters were shown to play a decisive role. We found that dynamic response and steady-state parameters for discrete systems are analogous to those for analog systems. Two conventional digital control system design tools, root-locus and Bode plots, were discussed, methods of construction were indicated, and ways to use these tools in design problems were illustrated. Section 4.4 on compensators showed that phase-lag and phase-lead compensators are used for frequency response modification. PID compensators were discussed as were deadbeat controllers. In the next chapter we discuss compensators as special kinds of digital filters and show how DSPs can be used to implement digital filters.

In this chapter we stress the importance of CAE software packages in the design of control systems. The Control System Toolbox from MATLAB provides several functions that greatly improve the efficiency of the design and analysis of control systems (see Appendix A). Key functions in the toolbox include RLOCUS, ZGRID, DBODE, and DSTEP. These are featured in discrete systems design. The employment of CAE tools renders the trial-and-error design of digital control systems with classical techniques a much less tedious task than it used to be a decade or two ago.

PROBLEMS

4.1 For the system of Fig. 4.47, find:

(a) The natural frequency ω_n

(b) The damping ratio ξ

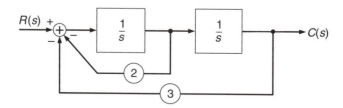

Figure 4.47   System for Problem 4.1.

4.2 For the system of Fig. 4.48, assuming that $R(s) = \frac{1}{s}$ find:

(a) The damping ratio

(b) The settling time of the system

(c) The rise time of the system

(d) The percent overshoot

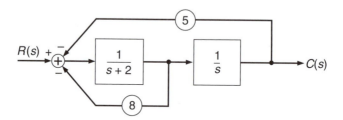

Figure 4.48   System for Problem 4.2.

4.3 Determine the position error constant, k_p, and the steady-state error for the system of Fig. 4.49.

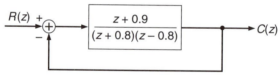

Figure 4.49   Discrete-time system.

4.4 Determine the velocity error constant, k_v, and the steady-state error for the system of Fig. 4.50.

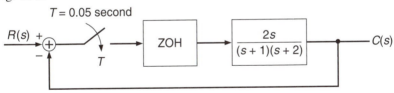

Figure 4.50   Discrete-time system for Problem 4.4.

4.5 For the sampled-data servo system depicted in Fig. 4.51, construct the root locus of the system in the z-domain as k varies from zero to infinity, after converting the system to its z-domain equivalent form. Use CAE tools of your choice.

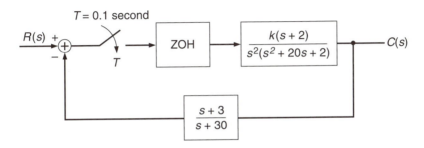

Figure 4.51   System for Problem 4.5.

4.6 Find the value of K in the control system of Fig. 4.52 such that the steady-state error for a ramp input is less than or equal to 0.05.

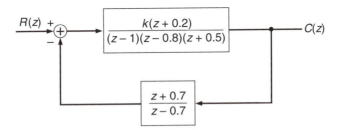

Figure 4.52   System for Problem 4.6.

4.7 In the system of Fig. 4.53, find the value of K and the value α such that the system has 20 % overshoot to a step input and a steady-state error of less than 0.02.

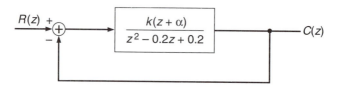

Figure 4.53   System for Problem 4.7.

4.8 Use the root-locus method to find a range of values for α so that the system in Fig. 4.54 is stable.

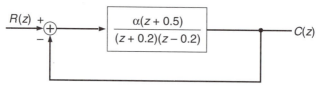

Figure 4.54   System for Problem 4.8.

4.9 For the system of Problem 4.8 construct the root-locus plot using MATLAB and locate the point in the z-plane and the value of α where the system has a damping ratio of 0.8.

4.10 Using the DBODE statement from MATLAB, determine the phase and gain margins for the system in Problem 4.9 at the value of α determined in that problem.

4.11 Design a phase-lead controller for the system in Fig. 4.55 assuming that we want a gain margin of 20 dB and a phase margin of 40 degrees.

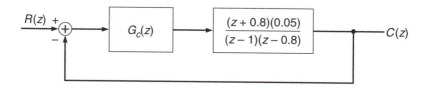

Figure 4.55   System for Problem 4.11.

4.12 For the system of Fig. 4.56 with sampling rate T construct a root locus as a function of T.

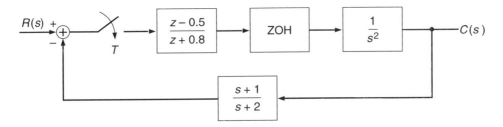

Figure 4.56 System for Problem 4.12.

4.13 Design a PID controller for the discrete-time system of Problem 4.11 meeting the same specifications. Compare results.

4.14 If the controller in Problems 4.11 and 4.13 were just a gain factor, what value should it be in order to guarantee a steady-state error of less than 0.1 for a ramp input?

4.15 Design a deadbeat controller for the system in Fig. 4.57.

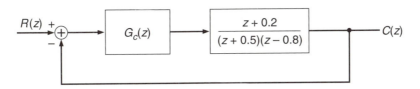

Figure 4.57 System for Problem 4.15.

4.16 Using the w-plane construct the Bode plot of the open-loop transfer function of the system in Fig. 4.58. Compare gain and phase margins with the plots from the MATLAB DBODE statement. Assume that T = 0.02 second.

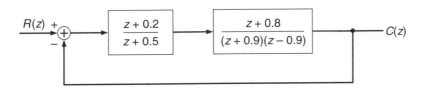

Figure 4.58 System for Problem 4.16.

4.17 For a sampled data system with zero order hold and unity gain feedback, if

$$G(s) = \frac{5}{s(s+2)}$$

and the closed-loop transfer function is

$$T(z) = \frac{0.02z + 0.02}{z^2 - 1.77z + 0.86}$$

determine the sampling frequency.

4.18   Construct the Bode plot of the unity gain feedback system with

$$G_p(z) = \frac{Kz(z - 0.5)}{(z - 1)(z - 0.9)(z - 0.8)}$$

using the MATLAB expression DBODE and determine the value of K such that
the system has a 45 degree phase margin.   Then what are the resulting system
steady-state errors to a step and ramp input?

REFERENCES

1.   C. L. Philips and H. T. Nagle, *Digital Control Systems Analysis and Design*, 2nd ed.,
     Prentice Hall, Upper Saddle River, NJ, 1990.

2.   H. F. Van Landingham, *Introduction to Digital Control Systems*, Macmillan, New
     York, 1985.

3.   G. F. Franklin, J. D. Powell and M. L. Workman, *Digital Control of Dynamic Systems*, 2nd ed., Addison-Wesley, Reading, MA, 1990.

4.   R. C. Dorf, *Modern Control Systems*, Addison-Wesley, Reading, MA, 1980.

5.   C. T. Chen, *Linear System Theory and Design*, CBS College Publishing, New York,
     1984.

6.   J. A. Cadzow and H. R. Martens, *Discrete-Time and Computer Control Systems*,
     Prentice Hall, Upper Saddle River, NJ, 1970.

7.   C.L. Philips and R.D. Harbor, *Feedback Control Systems*, Prentice Hall, Upper Saddle River, NJ, 1988.

8.   *Digital Control Applications with the TMS 320 Family: Selected Application Notes*,
     Applications Book, Texas Instruments, Dallas, TX, 1991.

9. C. Slivinsky and J. Borninski, *Control System Compensation and Implementation with the TMS32010 Digital Signal Processing Applications with the TMS320 Family: Theory, Algorithms, and Implementations; Application Book Vol. 1,* Texas Instruments, Dallas, TX, 1989.

10. *Digital Control System Design with the ADSP-2100 Family* Application Note, Analog Devices, Inc., Norwood, MA.

11. *Control Theory, Implementation, and Applications with the TMS320* Family Seminar Workbook, Texas Instruments, Dallas, TX, 1991.

12. B. C. Kuo, *Digital Control Systems, 2nd ed.,* Saunders College Publishing, Orlando, FL, 1992.

CHAPTER **5**

DSPs in Control Systems

5.1 INTRODUCTION

To select and implement a DSP as a microcontroller, we must understand the fundamentals of DSPs. This chapter presents basics of digital signal processors and their applications in control systems. Appendixes A, B, F, G, and H should be used in conjunction with this chapter:

Section 5.2 covers some of the key characteristics of DSP systems. It defines sampling theorem and compares analog with digital signal processing. In Section 5.3 we concentrate on single-chip DSPs. DSPs are compared with general-purpose microprocessors. A typical architecture of a DSP is examined and some of the architectural considerations when selecting DSPs are presented. We also compare fixed- and floating-point arithmetic types. Software tools for programming DSPs are discussed. Other DSP solutions, such as DSP cores, chip sets, and multiprocessors, are also discussed. Section 5.4 provides an overview of how DSPs are used in several control system applications.

Fig. 5.1 shows the basic flow of information in a typical DSP control design. As we discussed in earlier chapters, design and analysis are performed using conventional and modern techniques. (Chapter 6 covers modern design techniques.)

In the implementation phase we convert the design information, such as transfer functions or difference equations, into DSP algorithms. These algorithms are then converted into executable files used for programming DSPs. Usually, assembly languages are used for programming DSPs. In the experiment phase we insert the programmed DSP in the control system as the microcontroller, controlling/commanding the system, and start experimenting.

Verification of the DSP consists of simulation of the instruction sets, debugging of programs, and hardware emulation. Section 5.3 covers these steps in detail.

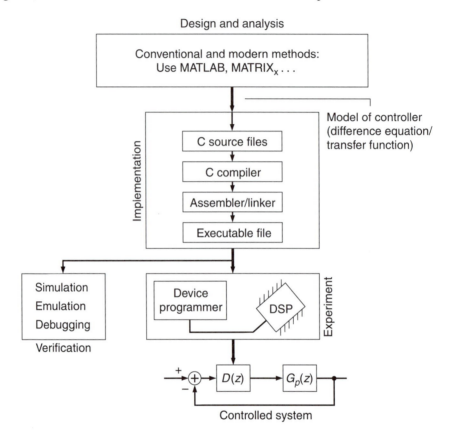

Figure 5.1   Process overview of DSPs used in control systems.

5.2 FUNDAMENTALS OF DIGITAL SIGNAL PROCESSING

Digital signal processing is defined as the arithmetic processing of signals sampled at regular intervals. Examples of this processing are filtering, convolution, amplification, modulation, and transformation of signals. In digital signal processing, we first use an ADC to convert continuous signals into discrete signals. Signal processing is performed on these digital signals, and finally, the processed digital signals are converted back to continuous signals using a DAC (see Section 3.2 for a discussion of the function of ADCs and DACs).

Fig. 5.2 shows this process. In this figure, low-pass filters are used for bandlimiting the input signals and for removing the effects of digitizing.

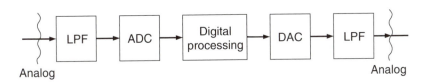

Figure 5.2   DSP operation.

Mathematical tools used in signal processing are Laplace transformation (in the continuous domain), z-transformation (in the discrete domain), and Fourier transformation (in the frequency domain). Some of the key characteristics of DSP systems are:

1. *Algorithms.* DSP mathematical operations are specified in algorithms and in turn these algorithms could be implemented on machine-language microcodes for a variety of target processors. Table 5.1 lists common DSP algorithms and their typical applications.

2. *Sample rates.* A unique characteristic of DSPs is sample rate. The sample rate in a DSP system is usually referred to the rate at which samples are produced from a continuous source.

Theorem 5.1 (Sampling Theorem)
If f_{max} is the highest frequency component of the system's frequency spectrum, and f_s is the sampling frequency, then for effective operation of a DSP system

$$f_s \geq 2f_{max} \qquad\qquad (5.1)$$

must be satisfied. In other words, the sampling rate must be at least twice the bandwidth or at least twice the highest-frequency component in the system. Theorem 5.1 is the basic signal-processing rule. It says essentially that if Equation 5.1 is not satisfied, information from the continuous domain will be lost in the process of sampling data from the continuous domain into the discrete domain. This means that the original signal cannot be reconstructed without introducing distortion. If sampling occurs at less than $2f_{max}$, we face an aliasing problem. *Aliasing* is basically the overlapping of the spectra or the introduction of noise into the desired signal passband. In control systems, the sampling rate selection is usually about 10 to 20 times the system's bandwidth.

Table 5.1   Common DSP Algorithms and Typical Applications

DSP Algorithm	System Application
Speech coding and decoding	Digital cellular telephones, personal communications systems, digital cordless telephones, multimedia computers, secure communications
Speech encryption and decryption	Digital cellular telephones, personal communications systems, digital cordless telephones, secure communications
Speech recognition	Advanced user interfaces, multimedia workstations, robotics, automotive applications, digital cellular telephones, personal communications systems, and digital cordless telephones.
Speech synthesis	Multimedia PCs, advanced user interfaces
Speaker identification	Security, multimedia workstations, advanced user interfaces
Hi-fi audio encoding and decoding	Consumer audio, consumer video, digital audio broadcast, professional audio, multimedia computers
Modem algorithms	Digital cellular telephones, personal communications systems, digital cordless telephones, digital audio broadcast, digital signaling on cable TV, multimedia computers, wireless computing, navigation, data/facsimile modems, secure communications
Noise cancellation	Professional audio, advanced vehicular audio, industrial applications
Audio equalization	Consumer audio, professional audio, advanced vehicular audio, music
Ambient acoustics emulation	Consumer audio, professional audio, advanced vehicular audio, music
Audio mixing and editing	Professional audio, music, multimedia computers
Sound synthesis	Professional audio, music, multimedia computers, advanced user interfaces
Vision	Security, multimedia computers, advanced user interfaces, instrumentation, robotics, navigation
Image compression and decompression	Digital photography, digital video, multimedia computers, video-over-voice, consumer video
Image compositing	Multimedia computers, consumer video, advanced user interfaces, navigation
Beamforming	Navigation, medial imaging, radar/sonar, signals intelligence
Echo cancellation	Speakerphones, modems, telephone switches
Spectral estimation	Signals intelligence, radar/sonar, professional audio, music

Source: © Berkeley Design Technology, Inc. (BDTI). See Reference [26].

3. *Clock Rate*. As with other synchronous devices and systems, DSPs have a clock rate that refers to the rate at which the device or system performs its basic unit of work. The ratio of system clock rate to sample rate is a major factor in determining

how a system is implemented. The required hardware to implement DSP algorithms depends on this ratio. For a higher clock system to sample rate ratio, a more complex hardware is required to implement the algorithms.

Analog versus Digital Signal Processing

Analog signal processing is based on the transfer function of a system. The component values of the system's analog circuit determine its transfer function and, therefore, its signal-processing performance. Although analog signal processing is more familiar to designers, easier to use, and provides high bandwidth, it is not efficient in today's signal-processing applications. The very limited number of purely analog applications and the large number of components in an analog signal processor are two reasons for the decline in popularity of analog signal processing. Because DSPs use software to control signal processing, they are more versatile than analog signal processors. Because of this versatility, DSPs are very useful in the design of digital filters. Table 5.2 compares analog and digital signal processors.

Table 5.2 Analog and Digital Signal Processing Compared

Type of Processing	Noise Immunity	Program-mable	Ease of Filter Implementation	Number of Components	Number of Applications	Resolution
Analog Signal Processing	No	No	Difficult	Large	Small	Low
Digital Signal Processing	Yes	Yes	Easy	Small	Large	High

5.3 SINGLE-CHIP DSPS

Because of a DSP's special architecture, it is more useful than a general-purpose microprocessor for high-speed processing applications. Table 5.3 shows the differences between single-chip DSPs and general-purpose microprocessors. DSPs have irregular instruction sets and are difficult to program. Because a control system is a real-time system, DSP architecture must handle a control system's numerical tasks and bandwidth requirements. Microcontrollers traditionally have a *von Neumann architecture* (meaning that instructions and data are on the same bus; see Fig. 5.3). However, most DSPs use a *Harvard architecture* (meaning that the data is separated from the instruction bus to increase speed) or a modified Harvard architecture that is optimized for signal processing.

Table 5.3   DSPs and General Purpose Microprocessors Compared

Function	Single Chip DSP	General Purpose Microprocessor
High speed math	Faster	Slower
Cost	High	Low
Interface with the outside world	Limited peripherals	On-Chip peripherals
Sampling rate	High	Low

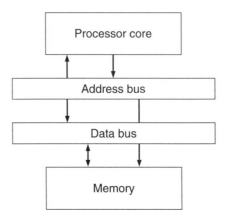

Figure 5.3   Von Neumann architecture. © Berkeley Design Technology, Inc. (BDTI). See Reference [26].

Fig. 5.4 shows a basic Harvard architecture. In this architecture, the processor can simultaneously access the two main memory banks using two independent sets of buses. Hence in this type of architecture two memory accesses can be made during any one instruction cycle.

The basic parts of a DSP architecture are the MAC (multiply and accumulate) unit, the program control unit, and the data/address bus interface units. Fig. 5.5 shows a typical DSP core. The integrated parallel hardware multiplier is the most distinguishing unit in this core. The MAC unit usually provides multiplication, multiplication with cumulative addition, and multiplication with cumulative subtraction.

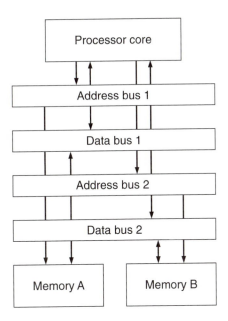

Figure 5.4   Harvard architecture.   © Berkeley Design Technology, Inc. (BDTI).   See Reference [26].

Fig. 5.5 shows that the two 16-bit data buses are input to the multiplier. The ALU (arithmetic logic unit) gets operands from:

- The accumulator through a feedback loop
- Memory
- The multiplier in the form of 32 bits

The ALU provides both arithmetic (add, subtract, negate, increment, and decrement) and logic (AND, OR, XOR, and NOT) functions.

You should select a DSP depending on the type of application. The most important considerations involve a DSP's architecture. Some of these considerations are:

- General arithmetic architecture.
- Ability to perform fast arithmetic and many other operations in one cycle.
- Circular buffering capabilities.
- Amount of address space.
- Unconstrained data flow to and from the arithmetic unit.
- Single-cycle instruction for high sampling rates. (This feature is possible because most functions are hard-wired logic within the architecture.)

- Multiple buses for simultaneous access of data and instructions.

- Hardware multiplier to minimize computational delays and for high-bandwidth applications for very fast sampling rates. Hardware multipliers traditionally were separate chips connected to DSPs. However, in modern DSPs, a hardware multiplier is part of the DSP architecture.

- Hardware shifters for fast scaling and large dynamic range.

- 32-bit registers to minimize truncation errors.

- At least 16-bit word length to minimize quantization errors.

- Interrupt capabilities (switching the program that is executing from one task to another).

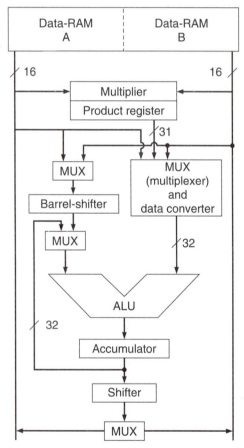

Figure 5.5   Typical DSP core. Reprinted with permission from Pergamon Press, Inc. See Reference [32].

Performance is usually measured in the number of cycles in which a DSP can accomplish the calculations above. Table 5.4 summarizes some of the architectural features of DSPs and their corresponding benefits.

Table 5.4 Architectural Features of DSPs

Feature	Benefit
Single-cycle instruction	Executes advanced control algorithms in real-time
Pipelined architecture	Controls high-bandwidth systems
Harvard architecture	Accesses data and instructions simultaneously
Hardware multiplier	Minimizes computational delays
Hardware shifters	Have larger dynamic range
Hardware stack	Supports fast interrupt processing
Saturation mode	Prevents wraparound of accumulator
32-bit registers	Minimizes truncation errors

Data and instructions are stored in RAM. Therefore, in many applications high-speed RAMs are required. On-chip as well as off-chip RAMs are available for different DSPs. Fast access time for these RAMs are key features of DSPs.

Fig. 5.6 shows the internal architecture of an ADSP-2101/2, which has a modified Harvard architecture. The ADSP-2101/2 architecture has three computational units (the arithmetic/logic unit, a multiplier/accumulator, and a barrel shifter), two data address generators, a program sequencer, a timer, interrupt capabilities, and on-chip program and data memory (all RAM). Because of its architecture, the ADSP-2101/2 can compute the following in a single cycle:

- Generate the next program address.
- Update one or two data address pointers.
- Receive and transmit data via the two serial ports.
- Fetch the next instruction.

Because intensive mathematical operations are required for typical control system applications, some of these operations are benchmarked by DSP chip manufacturers. Typical benchmarked operations are:

- Matrix multiplication
- Matrix inversion
- Real number multiplication
- Complex multiplication
- Boolean logic operations

- PID loop calculations
- Nth-order power series

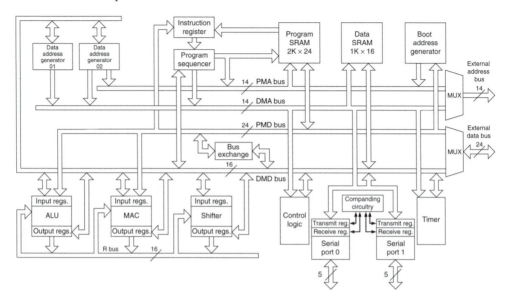

Figure 5.6   ADSP-2101/2 internal architecture. (Courtesy of Analog Devices, Inc.) See Reference [1].

Clock signals are used to sequence operations in DSPs. A DSP's master clock is the highest-frequency component clock and in some cases it comes from an external clock signal. On-chip frequency synthesizers are available on some DSPs and they generate a master clock from a lower-frequency input clock.

Transmitting and receiving data one bit at a time is done by the serial interface ports. Most commercially available DSPs have synchronous serial ports. This type of serial port transmits a one-bit clock signal along with the serial data bits. Serial ports are used for communicating with external peripherals. Unlike serial ports, parallel ports transmit and receive more than one bit of data at a time. As a result, parallel ports transfer data faster than serial ports. However, additional pins are needed on the device for this purpose. In addition to serial and parallel ports some DSPs provide a host port. A host port is usually a specialized 8- or 16-bits bi-directional parallel port that communicates with other DSPs and microprocessors. External Interrupt lines are pins on a DSP that can be used by an outside device to interrupt the processor.

The basic resolution and dynamic range of a DSP are functions of its word length. Fixed-point processors have a small dynamic range compared with the dynamic range of floating-point processors. In a fixed-point processor, both the signal and coefficients have to be represented in a finite amount of storage. Because of their finite word length and subsequent dynamic range limitations, fixed-point processors are susceptible to noise, and their registers overflow easily. These susceptibilities can cause instability. To reduce the finite-word length effects, the signals and their coefficients are scaled with a scaling factor.

Table 5.5 shows the differences between fixed- and floating-points arithmetic types. The main difference between fixed- and floating-point arithmetics is how the result of multiplication operation is used.

Table 5.5   Fixed- and Floating-Point Arithmetic Types Compared

Type of Processor	Finite Word-Length Effects	Cost	Scaling	Speed	Chip Estate	Word Length	Programming Tools	Dynamic Range	Resolution
Fixed-Point	Yes	Lower	Yes	Lower	Large	Small	No	Small	Low
Floating-Point	No	Higher	No	Higher	Small	Large	Yes	Large	High

Commercial fixed-point DSP processors are in the form of 16, 20, and 24-bit representation. Floating-point processors have the form of 32-bit processors. Even though fixed-point processors are difficult to work with because of their low word length, they are used in a lot of control applications. Some of these are robotic joint control, disk-drive actuators, high-speed servo systems, magnetic levitation systems, active suspension of vehicles, and platform stabilization. On the other hand, floating-point DSPs are ideal for projects where rapid prototyping is needed. Floating-point DSPs are used in robot coordinate transformation, high-order system identification, and complex nonlinear continuous systems. Tables 5.6 and 5.7 show the capabilities of popular fixed- and floating-point DSPs. To obtain the latest information on these DSPs and others, contact the manufacturers. Other considerations in selecting DSPs are the availability of software development tools for debugging, interfaces to external devices, and cost.

Table 5.6   Capabilities of Popular Fixed-Point DSPs

Company	Model Number	Word Length (Bits)	Clock (MHz)	Pins, Package	Miscellaneous
Analog Devices, Inc.	ADSP-2171	16	33	128-lead TQFP/PQFP	1. 24-bit instruction word, 14-bit data word 2. 2K x 14-bit data RAM 3. 4K x 24-bit program RAM 4. 8K x 24-bit program ROM

Lucent Tech- nologies, Inc.	DSP1617	16	50/30	100-pin PQFP	1. Four 36-bit accumulators 2. 24-Kbyte ROM 3. 4-Kbyte dual-ported RAM 4. MAC unit with two 36-bit accumulators 5. Two 64-Kbyte address spaces.
Motorola, Inc.	DSP56156	16	60/40	112-pin CQFP	1. 4-Kbyte program RAM or 24-Kbyte program ROM and 4-Kbyte RAM 2. Two 128-Kbyte address spaces 3. External bus: 16-bit address, data MAC 4. 1-cycle MAC
	DSP56001	24	33/27	132-Pin PQFP	1. 24-bit data, instruction; 16-bit address 2. 24 x 24-bit MPY, 56-bit accumulator 3. Two address spaces: 192-Kbyte program, 304-Kbyte data 4. External bus: 16-bit address, 24-bit data 5. 1-cycle MAC
Texas Instru- ments, Inc.	TMS320C14	16	25.6	44-Pin PLCC	1. 16-bit instruction, data 2. Two 16-bit registers, 32-bit accumulator 3. 512-byte RAM 4. 4-Kbyte ROM/ EPROM 5. 4 K-word address space
	TMS320C24	16	25.6	132 PQFP	1. 16-bit instruction, data 2. 16-bit T registers, 32-bit P register, 32-bit accumulator 3. 544 words x 16 of RAM 4. 16K words x 16 of ROM or Flash memory 5. 4K-word address space

Texas Instru- ments, Inc.	TMS320C28	16	40	80-Pin QFP 68-Pin PLCC	1. 2. 3. 4. 5.	64-byte data RAM. 16-Kbyte program ROM Two 64K-word address spaces 1-cycle instruction execution 1-cycle MPY, MAC
	TMS320C50	16	40/57/80	132-Pin PQFP	1. 2. 3. 4. 5.	20-Kbyte program/data RAM 4-Kbyte ROM Single-cycle MPY 1-cycle instruction execution Parallel ALU, MPY, and logic operations

Table 5.7   Capabilities of Popular Floating-Point DSPs

Company	Model Number	Word Length (Bits)	Clock (MHz)	Pins, Package	Miscellaneous
Analog Devices, Inc.	ADSP-21020	32	33	223-Pin PGA	1. 48-bit instruction, 32/40 bit data words 2. 24-bit program, 32-bit data address spaces, memory buses 3. 80-bit MAC accumulator
Lucent Technologies, Inc.	DSP3210	32	55/66	132-pin PQFP	1. 32-bit address space 2. 32-bit barrel shifted 3. 1-Cycle MAC 4. 1-Kbyte boot ROM 5. Two 4-Kbyte RAM blocks
Motorola, Inc.	DSP96002	32	33.3/40	223-Pin PGA	1. 32-bit address registers 2. 96-bit accumulator 3. Two 2-Kbyte RAM blocks 4. Two preprogrammed data ROMs 5. 1-Cycle MAC, instruction execution
Texas Instruments Inc.	TMS320C31	32	27/33/40	132-Pin PQFP	1. 32-bit data, instruction 2. 32-bit barrel shifter/ALU 3. Two 4-Kbyte data RAM blocks 4. 24-bit address space 5. 1-cycle MPY, ALU operations
	TMS320C40	32	40/50/60/80	325 Pin PGA	1. Two 4-Kbyte dual-access RAM blocks 2. Two external buses: 31-bit address and 32-bit data 3. Circular, bit-reversal addressing

Development Tools

Fig. 5.7 shows typical tools available for development of DSPs as microcontrollers.

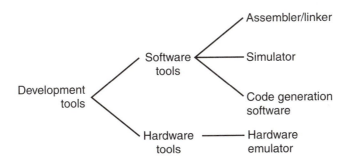

Figure 5.7   Development tools for DSPs.

High-level languages such as C and Pascal are not efficient for DSP programming. DSPs are usually programmed using assembly language.

The need for a universal signal-processing language written specifically for DSPs is apparent. A signal-processing language that currently exists is the DSPL language by dSPACE GmbH. DSPL is independent of the target hardware system and has extensive support for fixed-point processors. For example, it has automatic scaling. DSPL has been used in DSP programming of control system applications such as robotics, active suspension controllers, and filters.

Because of irregular instruction sets, programming DSPs is difficult and time consuming. Assembler/linkers and software simulators are also available to develop and debug DSP algorithms from all DSP vendors. An assembler translates assembly language source files into machine language object files for a specific DSP. Macro assemblers allow DSP programmers to use parameterized blocks of code. The linker combines multiple object files into a single executable object file. Debugger is the front-end user interface program. Different types of DSP debugger are graphical, source-level, signal plotting, character-based, and watch variables.

Software simulators are used for simulation of instruction sets of the device. Simulation speed is on the order of thousands of instructions per second. You can usually perform the following tasks with a software simulator:

- Debugging without the target hardware

- Modification of registers

- Initialization of memory before a program is loaded

- Manipulation of register and memory contents

These simulators are provided by the DSP vendors. However, there are differences in speed, accuracy, and debugging support between these simulators. As a result, the designer should investigate software simulation capabilities for a DSP before choosing one.

Hardware emulators are used for system-level integration and debugging. Hardware emulators allow the user to perform full-speed emulation with real-time hardware break-points. These emulators are normally used in systems with a single DSP.

Development boards are used for device evaluation and debugging. Real-time signal processing such as to write and run source code on the target DSP can be accomplished.

Code generation software automatically generates assembly code for a particular DSP from a description of the control algorithm. Examples of this type of software are Impex from dSPACE (see Appendix B for details on this product) for the TMS320 and DFDP from Atlanta Signal Processors, Inc.

Fig. 5.8 shows the TMS320 development product integration. In this figure, the C compilers translate C-language files into TMS320 assembly language source files. The assembler creates COFF (common object file format) and the linker translates these files into a single executable object file. The linker also takes archive library modules from a previous linker run. The role of archiver is to collect a group of files into a single file. The TMS320 C compilers, macro assemblers, and linkers are available for both fixed- and floating-point devices.

The software simulators take the object code from the macro assembler/linker or C compiler. EPROM programmers take the converted version of the COFF object file (usually in Intel or TI-tagged hex object format) as their input.

Up to this point we have discussed the most widely used form of DSP, a single-chip DSP. Other forms of DSPs are also used for different applications. Some of these forms are:

- *DSP cores.* A DSP core is the basic DSP building block in a larger chip. Therefore, the DSP core is not an off-the-shelf chip. Fig. 5.9 shows the basic architecture of a DSP-core ASIC. Examples of DSP cores are T320C2XLP and TEC320C52 from Texas Instruments and pine and oak DSP cores from DSP Group.

- *Multi-Processors (ASICs).* To achieve more performance and add more features on a DSP chip, some DSP vendors offer multi-processor ASICs. This type of ASIC contains a DSP and a microprocessor or a microcontroller. This form of DSP has more features than those of a single-chip DSP. However, only a few devices are commercially available.

- *Chip Sets.* Some DSP vendors have come up with separate packages for each specific function of a DSP. In this form the DSP functionality is spliced into multiple ICs. One advantage of chip sets over single-chip DSPs is that the designer has the flexibility of combing individual ICs in a form that best suits the specific application.

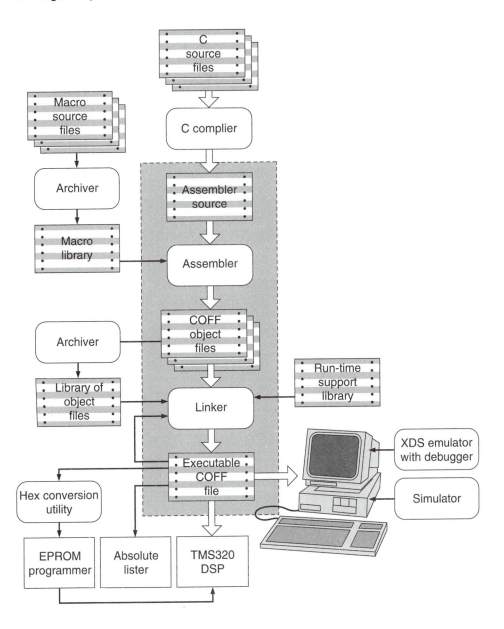

Figure 5.8   TMS320 development product integration. Courtesy of Texas Instruments. See Reference [29].

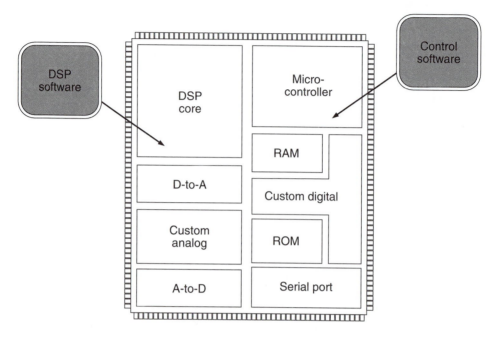

Figure 5.9   DSP-core ASIC. © Berkeley Design Technology, Inc. (BDTI). See Reference [26].

5.4 APPLICATIONS OF DSPS IN CONTROL SYSTEMS

In this section we give examples of how DSPs are used in several control system applications. The physical system is described briefly with its unique features and the type of controller/DSP used is also discussed.

PID Controllers

Since PID controllers are used as simple solutions in a variety of control system designs, we consider them first. In Section 4.3 we discussed PID controllers and their effects on control systems. PIDs are implemented as algorithms on both fixed- and floating-point DSPs. DSP manufacturers usually provide information on how to implement a PID algorithm that is suitable for their DSPs. Implementing a PID algorithm on a floating-point DSP is straightforward. However, for fixed-point DSPs, we must consider the following general guidelines:

- We need to know the order of magnitude of all variables for fixed-point calculations. Therefore, a prior knowledge of signals and parameter range is needed.
- Use available programs and simulators from the DSP manufacturer to test the effects of scaling and round-off.
- When selecting the sampling period for the controller, use

$$\frac{n}{T_I} \approx 0.2 \qquad (5.2)$$

for PI controllers, where n is the sampling period and T_I is the integral time. For PID controllers use

$$\frac{nN}{T_D} \approx 0.2 \text{ to } 0.6 \qquad (5.3)$$

where N is the maximum derivative gain and T_D is the derivative time.

- To avoid round-off errors, select a word length of at least

$$\text{Number of bits} = \frac{-\text{Log}(Kn/T_I)}{\text{Log}(2)} \qquad (5.4)$$

where K is the gain.

- You must scale the gain, input, and output variables.
- Interfacing of the DSP to the plant via an ADC and a DAC must be considered.

Motor Control

Before we cover applications of DSPs in motor control, we should summarize the basics of DC and AC motors.

DC motors

A DC motor converts current (usually generated by power amplifiers) into rotational torque. Fig. 5.10 shows a simple model of a DC motor. Some of the key parameters are k_t, the torque constant in units of Nm/A; r, the armature resistance in units of ohms; J_m, the moment of inertia in units of $kg - m^2$; and the maximum torque levels. The level of torque a motor can produce can be represented by a continuous value and a peak value. The continuous torque is a good measure of how the motor can produce torque continuously without overheating. The field and the armature currents are two independent controllable values.

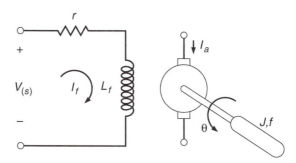

Figure 5.10   Model of a DC motor.

AC Motors

Induction motors are generally more reliable, less expensive, and need less maintenance than DC motors. However, they are nonlinear systems and require complex algorithms in their implementations. Also, AC induction motors are mostly used in constant-speed applications. In AC motors, the three phase currents are coupled together and are not independently controllable. Both torque-current and current-flux relationships are nonlinear.

Several methods are available for control of AC motors: vector control of induction motors and the principle of field-oriented control (FOC).

In these types of control, the magnetic flux is calculated and fed back to the control unit as a basis for the commutation of the stator current vector. Therefore, the torque-generating component of the stator current is decoupled from the field-producing component. As a result, the machine acts like a field-wound DC motor.

DSPs play a very active role in a variety of motor control applications. These include:

1. Implementation of simple DC motor solutions. For small DC motors the DSP can control motor position, velocity, and the torque. Conventionally inexpensive PID controllers are used for the control of DC motors.
2. Real-time generation of commands such as polynomials or lookup tables.
3. Integration of controllers, command generators, and signal conversions into one solution.
4. Implementation of advanced algorithms/controllers for AC induction motors. These controllers usually consist of one or more of the following: LQR, Kalman filters, notch filters, fuzzy logic and adaptive controllers.

AC induction motors require very advanced and complex techniques such as nonlinear and adaptive algorithms. See Reference 28 for implementation of a nonlinear linearization technique that is used to control the torque and flux of an induction motor. An observer (a Kalman filter in this case) and a nonlinear controller are used for implementation of the algorithm above. The heart of the controller is a TMS320C30 floating-point DSP from Texas Instruments, which operates from a dSPACE system.

Fig. 5.11 shows a block diagram of a DSP-based motor control solution. Here the command generation (such as polynomial and lookup tables), the digital controller (such as PID, LQR, adaptive controllers, and fuzzy logic), and the signal conversions (D/A and A/D) can all be implemented on a DSP controller.

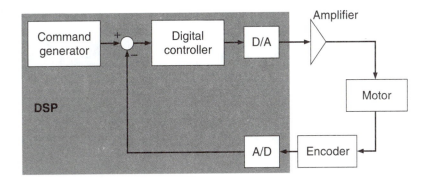

Figure 5.11   DSP integration of several components of a motor control system.

TI's TMS320C24x DSP is currently used for industrial applications (motor drives, power inverters, and motion controllers) and automotive control applications (electric power steering systems, antilock braking systems, and powertrain control systems). Fig. 5.12 Shows the key features of the TMS320C24x architecture. These include:

- 20-MIPS DSP core for real-time processing of advanced algorithms
- Optimized event manager
- Integrated dual 10-bit A/D converters
- SPI and SCI for serial communications
- 16K words of ROM or flash memory
- 28 bidirectional I/O pins
- Watching timer

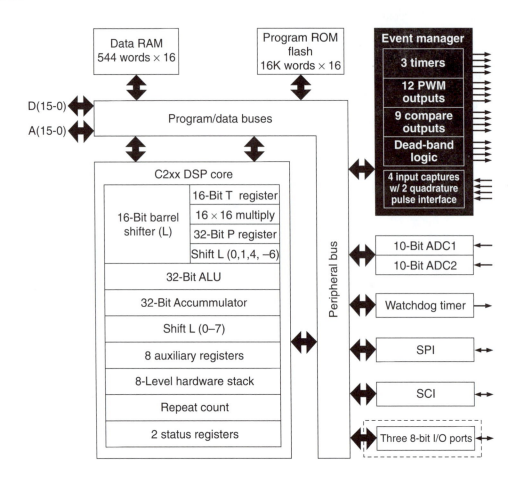

Figure 5.12   TMS320C24x DSP controller. Courtesy of Texas Instruments. See Reference [24].

Example 5.1[1]

ADSP-2115 DSP and the ADMC200 motion co-processor from Analog Devices are being used in high-speed AC motor control systems. Fig. 5.13 shows a field-oriented control system for a permanent-magnet synchronous motor. The system consists of a permanent magnet AC servomotor with a shaft-mounted resolver, a three-phase power inverter, and the

---

[1] This example is presented by courtesy of Analog Devices, Inc. Portions reprinted with permission. See Reference [27].

motor control circuit. The primary ICs in the control circuit are the ADSP-2115, the ADMC200, and the AD2S90 resolver-to-digital converter. The DSP is the shaft control processor and carries out all the motion control and torque current loop functions. The ADMC200 is the interface between the DSP and the inverter, and in addition provides the vector transformation functions required for AC motor control. The interface to the host controller can be either via the DSP data and address bus or via the serial port.

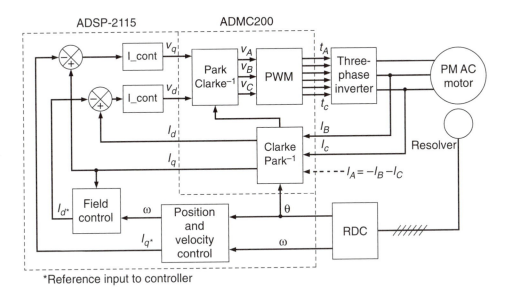

Figure 5.13   Motion control system.

The outer position and velocity loop calculates the torque demand, which is the input V_d reference for the current loops. At speeds less than the base speed, the $s = -1$ reference current will be zero. If an extended constant power speed range is required, the field control scheme can introduce some field weakening by setting a negative I_d^* value as a function of the motor speed.

The ADMC200 samples the motor currents and also transforms AC current waveforms into direct and quadrature current components within a rotating frame. A current loop control algorithm implemented on the DSP calculates desired V_d and V_q voltages for the motor. The ADMC200 forward vector transformation block also maps direct and quadrature motor voltages into AC voltages within the stator reference frame. The DSP scales and then write these results to the PWM block of the ADMC200.

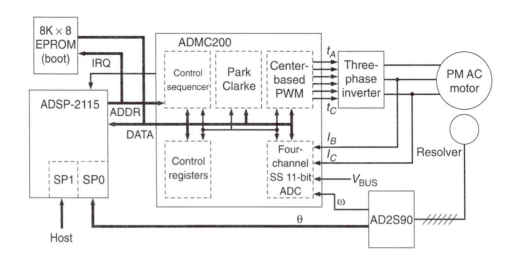

Figure 5.14   Overview of the system control hardware.

The motion control software is stored on an external 8-bit EPROM and is loaded automatically into the DSP's 1K words of internal program RAM on power-up. Each of the 24-bit program words is stored on the EPROM in a 4-byte segment. The DSP boot firmware copies the program from the EPROM to the internal RAM in the correct order to rebuild the 24-bit-wide program memory. This arrangement limits the external EPROM requirement to just a single slow memory device.

It is possible to switch eight pages of program memory RAM stored in a 32K EPROM. For example, the first boot page could contain the programs which will initialize all data variables (lookup tables, etc.), configure the ADMC200 registers (setup PWM registers, etc.), and perform self-diagnostic functions, while the second page can contain the motion control algorithms which are loaded at the end of the initialization phase.

The ADMC200 device can be connected directly to the DSP data and address buses. The internal registers can be written to in the same way as data RAM placed in the low memory address space. The shaft control algorithm can be timed through the ADMC200 CONVST pin or via the interrupt pin on the DSP.

Table 5.8 outlines the control algorithm. The functions in bold are implemented on the ADMC200 co-processor. The scheduling of the control algorithm is synchronized with the ADC interrupt service routines. In this application the ADC will generate an interrupt at a rate of 10 KHz. The torque control loop is realized at this frequency, while the motion loop is scheduled every fourth ADC sample. The basic torque loop functions can be carried out in less than 20 μs; this leaves the remaining 80% of the time for the motion control loop and other functions. A more detailed explanation of the torque control loop is given in Table 5.9.

Table 5.8 Control Algorithm Outline.

Inputs	ADMC200 Write Register	Motion_Control Functions	Out	ADMC200 Read Register
		Read_RDC	ω ρ	ADCAUX
ω		position_velocity_loop	I_q^*	
ω		Field_control $(I_d^* = 0$ for $\omega < \omega_{base})$	I_d^*	

		Torque_Control Functions		ADMC200 Register
		Sample_phase_currents	I_b I_c	ADCV ADCW
I_b I_c ρ	PHIP2/VQ PHIP3 RHO	**Clarke_Park^{-1}** $(I_a = -I_b - I_c)$	I_d I_q I_a	ID/PHV1/VX IQ/PHV2 IX/PHV3
I_q I_d^* ω		I_control_q	V_a	
I_d I_d^* ω		I_control_d	V_d	
V_q V_d ρ	PHIP1/VD PHIP2/VQ RHOP	**Park_Clarke^{-1}**	V_a V_b V_c	ID/PHV1 IQ/PHV2 IX/PHV3
V_a V_b^* V_c		PWM_scale	T_a T_b T_c	
T_a T_b T_c	PWMCHA PWMCHB PWMCHC	**PWM_out**		

Table 5.9 Torque Loop Algorithm.

Start_torque_loop	Wait for ADC interrupt	Elapsed time
Read_currents	ADC_int: read Iph(2) from ADMC200: ADCV read Iph(3) from ADMC200: ADCW	1.1 µs
load PARK registers for stator to rotor transformation	write Iph(2) to ADMC200: PHIP2 write Iph(3) to ADMC200: PHIP3 write ρ to ADMC200: RHO	
Clarke_Park^{-1}	**meanwhile** check for over current wait for RPARK interrupt	5.1 µs
read PARK registers	RPARK_int: read Id from ADMC200: ID read Iq from ADMC200: IQ read Iph(1) from ADMC200: IX	5.9 µs
I_control_d error driven PI loop (save DVq_n and DIq_n values) + machine equations	DId_n1 = Id_ref - Id DVd_n1 = DVd_n + KPd*(DId_n1-KId*DId_n) DVd_n = DVd_n1 DId_n = DId_n1 Vd = DVd_n1 + (L$_s$.I$_q$)*velocity + [IRd = I$_d$*.R$_s$]	7.8 µs
I_control_q: error driven PI loop (save DVq_n and DIq_n values) + machine equations	DIq_n1 = Iq_ref - Iq DVq_n1 = DVq_n + KPq*(DIq_n1-KIq*DIq_n) DVq_n = DVq_n1 DIq_n = DIq_n1 Vq = DVq_n1 + (K$_E$ + L$_s$.I$_d$)*velocity + [IRq = I$_q$*.R$_s$]	9.6 µs
load PARK registers for rotor to stator transformation	write Vq to ADMC200 VQ; write Vd to ADMC200 VD; write ρ to ADMC200: RHOP;	9.9 µs
Park_Clarke^{-1}	**meanwhile** do some velocity filtering wait for FPARK interrupt	13.9 µs
read PARK registers	FPARK_int: read Vph(1) from ADMC200: PHV1 read Vph(2) from ADMC200: PHV2 read Vph(3) from ADMC200: PHV3	14.8 µs

PWM_out
deadtime adjustment
reverse for -ve Iph
calculate PWM times
write to registers
	for I = 1,3
	  T0 = TPWM/2 + TPD
	  IF Iph(I) LT 0 T0 = TPWM/2 - TPD
	  T(I) = T0 + VSCALE*Vph(I)
	    write T(I) to PWMCH(I)_ADMC200
	end_for_loop
		18.0 µs

Sample assembly code for some of these functions is given in Appendix G. The current loop equations are based on a PI-driven error loop and the machine-winding model. The machine equations for the U_d and U_q voltages are

$$U_d = I_d R_s + \omega_r (L_s I_q)$$
$$U_q = I_q R_s + \omega_r (L_s I_q + K_E) \tag{5.5}$$

The PI loop is of the form

$$G_{PI}(s) = K_P (1 + \frac{K_I}{s}) \tag{5.6}$$

The discrete form of this can be obtained by substituting

$$s = \frac{T_s}{2} \left(\frac{z-1}{z+1} \right) \tag{5.7}$$

This gives an equation of the form

$$G_{PI}(z) = KP \left(\frac{z + KI}{z - 1} \right) \tag{5.8}$$

where

$$KP = K_P (1 + \frac{K_I T_s}{2}) \tag{5.9}$$

$$KI = \frac{1 - (\frac{K_I T_s}{2})}{1 + (\frac{K_I T_s}{2})} \tag{5.10}$$

The difference equation for this transfer function is

$$DV_{K+1} = DV_K + KPDI_{K+1} + KPKIDI_K \tag{5.11}$$

The applied V_d and V_q voltages are the sum of the calculated machine winding voltages, Equation 5.5, and the error-correcting term from the PI loop Equation 5.11:

$$V_d = \Delta V_d + U_d$$
$$V_q = \Delta V_q + U_q$$

(5.12)

In addition to the ADMC200 motion co-processor, we can also use the ADMC300 series of DSP-based servomotor controllers from Analog Devices. Some features of the ADMC300 are:

- 25 MIPS fixed-point DSP core
- Single cycle instruction execution (40ns)
- High-resolution multichannel ADC system
- Multifunction instruction
- Three-phase PWM generation subsystem
- Programmable interrupt controller
- Flexible encoder interface subsystem
- Two dedicated ADC interrupts
- Suitable for AC induction and synchronous motors
- Special crossover function for brushless DC motors

Motion Controllers

A motion controller is basically used in a control system (servo system) to decode the motor position feedback and to generate the desired position. Appendix F provides some industrial motion controllers. A single-board motion controller has a set of configurable parameters that let the designers control the desired features. These parameters can be set using a personal computer.

In motion control systems, we normally deal with high-order controllers and sensors. Therefore, as in robotics, we can use a DSP for the large number of computations per sampling step that is required. Example 6.14 shows several ways to use DSPs in a motion controller. Several control strategies, such as a velocity observer and a disturbance observer, are implemented on a motion control card (MCV60). The heart of the MCV60 hardware is a TMS320C25 processor.

Robotics

DSP techniques are used in the design and control of robots. The application of DSPs in robotics takes different forms. For example, a DSP-based system can be used as the controller to implement control methods for robotic hands. DSPs can also be used to stabilize structural vibrations in a high-acceleration robot.

Once the equations of motion are set up, we will know the system's parameters, and therefore we can design a controller for the overall system. The conventional controllers for robots are normally PID controllers. But with the advent of DSPs, modern controllers (Kalman filters and state-space methods) are implemented as a mathematical algorithm on a suitable DSP.

The following guidelines are summarized for the application of DSPs in robotics.

- For nonlinear robot control and inverse kinematics problems, use floating-point DSPs because of their large dynamic range.

- In fast robots, replace conventional controllers, such as PID controllers, with modern controllers such as state estimators, and implement the resulting algorithms on DSPs.

- Use a DSP as a controller to achieve high sampling rates (tens of kilohertz).

- Use a DSP as a single controller to control a system's variables, such as velocity, position, torque, and current.

- Use DSPs to implement notch filters for removing unnecessary vibrations in a robot.

- Use DSPs to implement adaptive filters/algorithms for universal controllers.

Example 5.2[2]

A multivariable controller for a compliant articulated robot has been implemented using the dSPACE tools. This controller provides steady-state accuracy of the tip position on high-acceleration trajectories and active damping of structural vibrations which are excited with fast-changing accelerations or external disturbances. Digital signal processor (DSP) performance contributes to achieve the required sampling rate. During recent years, increasing demands on the accuracy and speed of industrial robots increasingly required abandonment of standard controller concepts and implementation of more sophisticated high-bandwidth multivariable digital controllers. To damp or at least stabilize the structural vibrations that are excited with high-bandwidth controllers there is the need of high sampling rates in the kilohertz range.

Fig. 5.16 shows a six-joint articulated system that is controlled completely by dSPACE DSP hardware hosted in a standard PC/AT environment. Characteristics of the robot are DC disk armature motors with harmonic drive gears for each joint and thin-walled aluminum positioning arms to permit considerable elasticity to the system, with dominant, even visible vibrations in the frequency range of about 10 to 150 KHz. Additionally, the gears have Coulomb friction, which must be overcome by the motor torque.

---

[2] This example is presented by courtesy of dSPACE, GmbH. Portions reprinted with permission. See Reference [30].

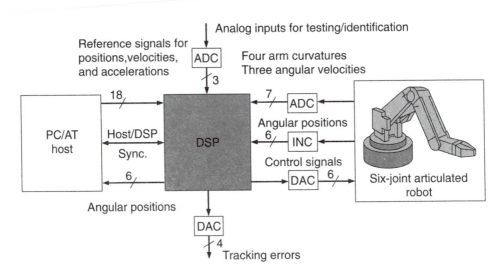

Figure 5.16   Robot vibration-damping and tracking control system.

With respect to high-acceleration and precise tip positioning control, these mechanical inconveniences require nonstandard control algorithms and control hardware.

The control algorithm has been designed as a measurement vector feedback and feed-forward of appropriate reference signals and has been implemented on the DS1001 processor board from the dSPACE DSP-CITpro hardware line with the TMS320C25 DSP. Using the dSPACE software tools the implementation of this controller required only a few minutes. Vibration damping and sufficiently high closed-loop bandwidth of the joint position loops are provided by the feedback control angular position and velocity from incremental encoders and tachometers, and nonstandard curvature measurements from wire straingauges attached to the positioning arms (vibration control). Steady-state accuracy of the tip position on desired tracks is achieved by appropriate feedforward of the reference positions, velocities, and accelerations of all joints. These signals are calculated by the PC/AT host off-line to the DS1001 dual-port memory without interfering with the DSP running the control algorithm. Synchronization of the transfer is also accomplished via the dual-port memory. Additional signals in Fig. 5.17 are for testing, analysis, and monitoring purposes.

The entire controller has 35 inputs and 17 outputs and is running at a sampling frequency of 8 kHz. This very high sampling rate is due not only to the TMS320C25's computing performance. With the large number of inputs and outputs the highly efficient coupling of the peripheral boards (ADC, DAC, and incremental encoder interface) to the DSP board via the PHS bus and the highly efficient DSP-host interfacing of the DSP-CITpro hardware line are crucial.

While experiments with standard controllers showed visible vibration and steady-state tracking errors, the DSP-based vibration control gives surprising results. Even "very fast" trajectories (fast-changing accelerations) can be tracked with high accuracy. The improvement is shown in Fig. 5.18.

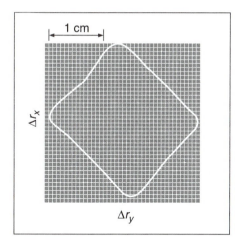

Figure 5.17   Standard control.

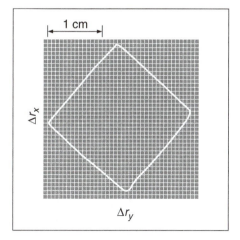

Figure 5.18   DSP control.

Disk-Drive Servo-Control

Fig. 5.19 shows the block diagram of a disk-drive servo-control system. The position of the read/write head on the disk is to be controlled using the servo-motor. High sampling rates are required for increased accuracy in head positioning and tracking. Also, control algorithms require intensive mathematical calculations. Therefore, DSPs are the preferred controllers for this type of control systems.

Traditionally, microcontrollers and DSPs together were used for the control/command unit in disk-drive systems. However, new DSP cores are being used to handle both the drive's functions (handled by microcontrollers) and positioning of the heads. This single-chip solution offers lower costs, improved reliability, and less power consumption.

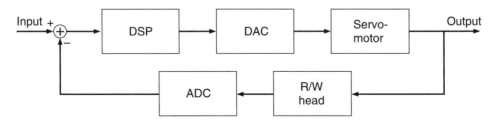

Figure 5.19   Disk-drive servo-control.

Example 5.3[3]

Fig. 5.20 shows a block diagram of a disk servo-control system. The drive input of a servo-motor that determines the position of the read/write head on the disk is the process to be controlled. The controller issues the appropriate commands to the servomotor via the D/A converter. The servomotor, in turn, moves the read/write head from a given position to any desired track on the disk.

The design objective is to keep the position error minimized at all times. A digital controller incorporates a filtering operation in its feedback loop, minimizing the position error and using the position error to control a motor via a DAC. Simultaneously, the processor must also maintain the disk speed at the desired level. To do so, the processor monitors a digital signal from a counter circuit that indicates the speed of the spindle motor and issues appropriate control commands to the spindle motor through a DAC.

Past and recent trends have indicated a consistent increase in the demand for larger disk storage capacities. Increasing the number of tracks on a disk while decreasing their width represents one method to meet this demand.

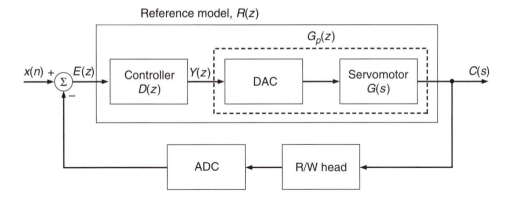

Figure 5.20   Block diagram of the servo-control system.

With this approach, however, accurate read/write head positioning and tracking becomes difficult with increasing disk capacity, thereby rending conventional design approaches inadequate. The demand for increased accuracy in head positioning and tracking requires a more frequent sampling of the head position than would otherwise be needed.

Moreover, this sampling rate must increase as the time constant of the process to be controlled decreases. Therefore, the disk controller must be capable of high sampling rates in addition to performing the math-intensive algorithms inherent in digital control applications. The required sampling frequency is governed by the desired transient response characteristics of the system (i.e., the closed-loop bandwidth). A general rule suggests a sampling rate that is 10 times the system bandwidth. The processor chosen to implement the controller must have the speed and word length necessary to achieve this sampling rate and properly implement the desired control algorithm.

Conventional design approaches give way to adaptive schemes when higher track densities demand increasingly accurate head positioning in the presence of environmental variations. Fig. 5.21 illustrates one such scheme. According to this scheme, the servo-plant output, c(n), must follow a reference (desired) model output, s(n). Those elements comprising the reference model appear in Fig. 5.22. A digital controller, $D(z)$, and servo plant, $G(z)$, comprise the reference model, $R(z)$, while a servomotor, $G(s)$, and DAC comprise the servo plant. The digital controller itself may be designed using conventional methods. Referring again to Fig. 5.21, the adaptive reference inverse model is an inverse model of the servo plant which when combined with the servo plant and reference model shown in Fig. 5.22 gives an output, y(n), that follows the reference model output, s(n). The adaptive reference inverse model is precomputed off-line. Once an adaptive reference inverse model is obtained, it is incorporated into the control system as shown in Fig. 5.21. The servo plant is driven by the output obtained from a copy of the adaptive reference inverse model are updated after each seek operation. This ensures that the servo plant follows the same profile as the reference model at all times. In the following section, the equations that describe the servo plant, reference model, and the adaptive reference inverse model are defined. To begin with, we assume that the servo-motor transfer function is given by

$$G(s) = \frac{1}{s(s+1)} \tag{5.13}$$

A general rule suggests that the sampling frequency can be chosen as one tenth of the fastest time constant. Since the time constant of the pole at $s = -1$ is 1 second, the time period, T, is chosen as 0.1 second.

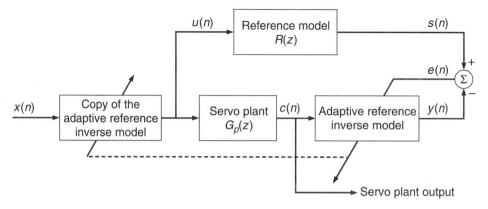

Figure 5.21   Model reference servo-control system.

Then, referring to Fig. 5.20, the transfer function of the servo plant is

$$G_p(s) = \big[G(s)\big]_{\text{[zero-order hold]}} \tag{5.14}$$

$$= \frac{1 - e^{-Ts}}{s} \frac{1}{s(s+1)} \tag{5.15}$$

$$= G_p(z) = \frac{C(z)}{Y(z)} = ZT\left[\frac{1 - e^{-Ts}}{s^2(s+1)}\right] \tag{5.16}$$

$$= \frac{z-1}{z} \, ZT \left[\frac{1}{s^2(s+1)} \right] \tag{5.17}$$

$$G_p(z) = \frac{0.004837z + 0.004678}{z^2 + 1.9z + 0.9094} \tag{5.18}$$

where ZT means "the z-transform of."

To achieve certain system characteristics, a digital controller (or compensator) needs to be designed. Several techniques exist for the design of digital compensators. The trade-offs involved in designing with different methods are explained here. With the frequency response procedures (Bode plot), an attempt is made to reshape the system open-loop frequency response to achieve certain stability margins, transient response characteristics, steady-state response characteristics, and so on. Thus the controller is designed with analog techniques, and a digital approximation is obtained with the standard z-transform, bilinear z-transform, or matched z-transform. A different design is obtained by the root locus. The root locus for a system is a plot of the roots of the system's characteristic equation as the gain is varied. The design procedure is to add poles and zeros via a digital controller to shift the roots of the characteristic equation to more appropriate locations in the z-plane. Both the phase-lag and phase-lead compensators allow a higher open-loop gain. The phase-lead controller results in a system with a smaller time constant; thus the system responds faster (larger bandwidth). As the conventional root-locus diagram allows only one parameter to vary at a given time, the design of a digital controller in the z-plane using a root-locus diagram is essentially a trial-and-error method.

For systems with more than one signal in the feedback path, pole-assignment and state estimation techniques can be used. The design results in the assignment of the poles of the closed-loop transfer function (zeros of the characteristic equation) to any desired location. This technique requires the measurement of all the states of the system if possible. If not, some states of the plant could be estimated from the information that is available concerning the plant. This results in the design of an observer or a state estimator. In pole-assignment technique, it is assumed that the pole locations that yield the best control system are known. However, another optimal design technique to obtain the best control system is to write a mathematical function called the cost function, resulting in an optimal control system. In this example, a phase-lead controller is designed using root-locus techniques. The transfer function of the controller is given by

$$D(z) = \frac{3.15(z^2 - 1.9z + 0.9094)}{z^2 - 1.6847z + 0.7147} \tag{5.19}$$

Now, the reference model transfer function is given as

$$R(z) = D(z)G_p(z) = \left[\frac{3.15(z^2 - 1.9z + 0.9094)}{z^2 - 1.6847z + 0.7147}\right]\left[\frac{0.004837z + 0.004678}{z^2 - 1.9z + 0.9094}\right] \quad (5.20)$$

or

$$R(z) = \frac{C(z)}{E(z)} = \frac{0.01524z + 0.0147}{z^2 - 1.68476z + 0.7147} \quad (5.21)$$

This reference model may now be used to derive the adaptive reference inverse model of the servo plant. Fig. 5.22 illustrates the adaptive reference inverse modeling technique. This particular model incorporates a 40-weight transversal filter whose coefficients are updated according to the least-mean-square (LMS) algorithm. The compromise between accuracy and computational complexity dictates the choice of the number of tapes in the transversal filter. The following equations describe this model:

•   Servo-plant output:

$$c(n) = 0.0048 * x(n-1) + 0.0046 * x(n-2) + 1.9 * c(n-1) - 0.9094 * c(n-2) \quad (5.22)$$

•   Reference model output:

$$s(n) = 0.01524 * x(n-1) + 0.0147 * x(n-2) + 1.68476 * s(n-1) - 0.7147 * s(n-2)$$
$$(5.23)$$

•   Adaptive reference inverse model output:

$$y(n) = \sum_{i=0}^{i=N} w(i) * c(n-i+1) \; ; \; N = 40 \quad (5.24)$$

$$\text{Error: } e(n) = s(n) - y(n) \quad (5.25)$$

•   Weight vector update:

$$W_i = W_i + \mu * e * c(n-i+1) \; ; \; i = 1,40 \quad (5.26)$$

Taking the inverse z-transform of $G_p(z)$ and $R(z)$, as given by Equations 5.18 and 5.21, respectively, yields the difference equations for the servo-plant output, c(n), and the reference model output, s(n). The weight vector, W_i, which represents the adaptive reference inverse model, is obtained by performing 2000 iterations of the adaptive loop. The parameter μ, which determines the rate of convergence for obtaining the weight vector, is chosen empirically, in this case, to be 0.05. The input to the system is assumed to be a step. Once the adaptive reference inverse model is obtained, it can be applied in the control system of Fig. 5.21. Due to the adaptive nature of the inverse model, any variations in the internal variables of the servo plant result in corresponding changes in the coefficients of the adaptive reference inverse model. Hence, the servo plant output follows the reference model at all times. The following equations describe the control system of Fig. 5.21.

- Servo-plant output:

$$c(n) = 0.0048 * u(n-1) + 0.0046 * u(n-2) + 1.9 * c(n-1) - 0.9094 * c(n-2)$$
$$(5.27)$$

- Reference model output:

$$s(n) = 0.01524 * u(n-1) + 0.0147 * u(n-2) + 1.68476 * s(n-1) - 0.7147 * s(n-2)$$
$$(5.28)$$

- Adaptive reference inverse model output:

$$y(n) = \sum_{i=0}^{i=N} w(i) * c(n - i + 1) \; ; \; N = 40 \qquad (5.29)$$

- Servo-plant input:

$$u(n) = \sum_{i=0}^{i=N} w(i) * x(n - i + 1) \; ; \; N = 40 \qquad (5.30)$$

$$\text{Error: } e(n) = s(n) - y(n) \qquad (5.31)$$

- Weight update:

$$W_i = W_i + \mu * e * c(n - i + 1) ; i = 1,40 \tag{5.32}$$

The DSP16/DSP16A adaptive control code, which appears in Appendix G, executes in 455 instruction cycles (11.37 μs on a 25-ns device). Fig. 5.23 shows the step response of the uncompensated system, while Fig. 5.24 shows the step response of the reference plant and the servo plant after adaptive compensation. These figures clearly show how closely the servo-plant output follows the reference plant output. In the case of a disk-drive system, the compensated system would allow much faster and accurate head positioning for read/write operations.

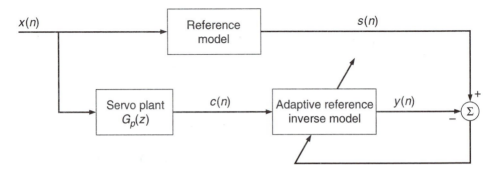

Figure 5.22   Adaptive inverse modeling scheme.

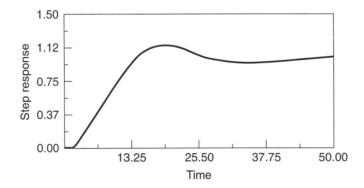

Figure 5.23   Step response of the uncompensated system.

Disk-Drive Controller Interface

Before reading and writing information on the storage medium, several disk operations must occur. First, the read or write location must be defined and the read/write head must be moved to the desired track. This operation is called a *seek*. Next, the analog signal representing the information must be digitized and applied to or extracted from the disk. Fig. 5.25 shows the hardware elements that effect these operations. Most disk drives have multiple platters for data storage. In such devices, track 0 of the first platter sits over track 0 of the second, third, and remaining platters. Thus particular tracks on all platters form spatial cylinders, so that selecting a track corresponds to positioning the read/write head over that cylinder. Specifying one of these multiple platters requires entering the drive via the controller (DSP16/DSP16A). Those commands that specify a given head cause the DSP to route selection signals over to the parallel bus. Once a particular head has been selected, control signals are sent through the DAC to the servo-motor. The demodulated data from the read/write head is read into the serial port through an ADC. At this point the DSP16/DSP16A computes the filter and sends the appropriate control signals to the servo-motor. Upon completion of the seek mode, the data are transferred from the disk to dual-port RAM. The converters used in this example are PCM78 (ADC) and PCM56P (DAC), which are 16-bit serial converters.

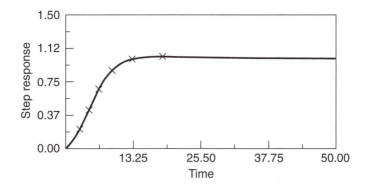

Figure 5.24   Step response of the compensated system. Crosses denote the response of the reference model.

Controller and CPU Interface

The need for a controller to have the capability to accommodate multiple host computer buses and attach a variety of storage devices led to the introduction of standard interfaces. An example of this is the small computers systems interface (SCSI). With this high-level interface for host computers, certain types of storage devices (disk, tape, floppy, optical disk, etc.) can be interchanged freely without requiring modification to the host system hardware or software. The attachment of this interface to a particular CPU bus requires a host adapter. Fig. 5.26 shows the block diagram of the controller host CPU interface. The data stored on the disk are encoded in one of several popular formats, including modified frequency-modulated (MFM) form. The data separator decodes the data on the disk into binary information that the CPU can understand. The sequencer block performs disk for-

matting and data read/write operations, as well as error correction and serializa-
tion/deserialization of the data from the host interface. The dual-port buffer allows for differences in speed between the host bus and the disk.

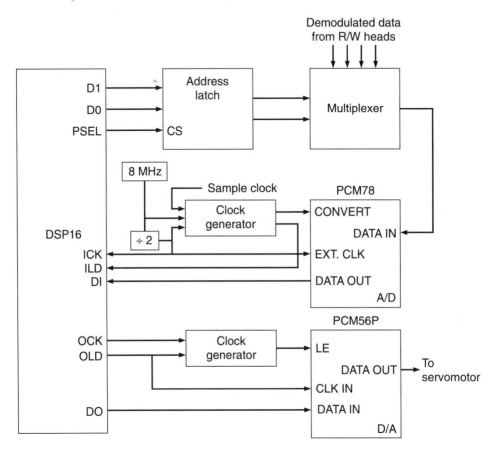

Figure 5.25   Controller and disk servo interface.

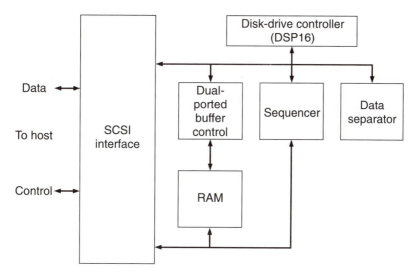

Fig. 5.26   Host controller interface.

Stabilization/Pointing Systems

DSPs are used in a servocontroller to stabilize a DC-motor driven, two-axis, gimbaled plat-form. Normally, in this type of system, the controllers have many notch filters as well as lowpass and loop-shaping filters.

Example 5.4[4]

The system used as an example of the application of the TMS32010[5] is a servo-control sys-tem for stabilizing a large, two-axis gimbaled platform with a DC-motor drive. Inertial rate-integrating gyroscopes mounted directly on the platform serve as angular motion sensors. Such systems are required for precise control of the line of sight (LOS) and line-of-sight rate for use in pointing and tracking applications for laser, video, inertial navigation, and radar systems.

At present, digital control is not normally used in systems of this type because of the fast throughput rates and computational accuracy required to perform the control computa-tions and notch filtering. Current line-of-sight stabilization systems continue to use analog electronics to implement servo-compensation functions and error-signal conditioning. Thus

---

[4] "Reprinted by permission of Texas Instruments." Portions reprinted with permission. Reprinted with permission from authors. See Reference [31].
[5] Texas Instruments has upgraded TMS32010 to TMS320C10; this example applies to both DSPs.

the system is representative in complexity and performance of typical systems currently in use by the aerospace industry and is a candidate for microprocessor-based digital control.

System model and control compensation. A single axis of the stabilization system has two primary control loops: the rate loop and the position loop. In addition, a tachometer loop exists within the position loop. The rate and position loops are identified in Fig. 5.27, a diagram of the elevation axis of the system. In its analog version, the system employs analog electronics to implement all control compensation and signal conditioning functions. Fig. 5.28 identifies those filters and compensators in the rate loop that are to be incorporated into the digital control system.

This study's approach provides a digital implementation of the designated analog elements of the rate loop without sacrificing closed-loop performance. Within the rate loop, the transfer functions to be implemented digitally consist of a first- and second-order compensatory, along with six notch filters. Within the position loop, there is one first-order compensatory and one notch filter. The transfer functions shown in Table 5.10, include both the analog prototypes and their digital equivalents.

The sampling rate chosen is 4020 samples per second (sampling period is 249 µs). This rate is more than twice the highest frequency of consequence (1800 Hz, the highest rate-loop notch frequency) to prevent aliasing. The rate is fast enough to prevent excessive phase lag in the rate loop and is more than 10 times the closed rate-loop bandwidth (approximately 80 Hz). The rate was also chosen to be an integer multiple of 30 Hz, which is a commonly used update rate of the video and infrared imaging/tracker devices that provide the line-of-sight rate command to the stabilization system's rate loop. The update rate of the imaging device and the sampling rate within the rate loop are thus synchronized.

After simulating the closed rate loop, the phase margin was found to be five degrees less than it was for the all-analog system, due to the computational and other delays associated with sampling. To overcome this deterioration in phase margin, the second-order rate loop compensator was redesigned to provide additional phase lead. The compensator was modified to provide enough additional phase lead so that the phase margin of the digital system matched that of the analog system. The modified compensator is listed in the table.

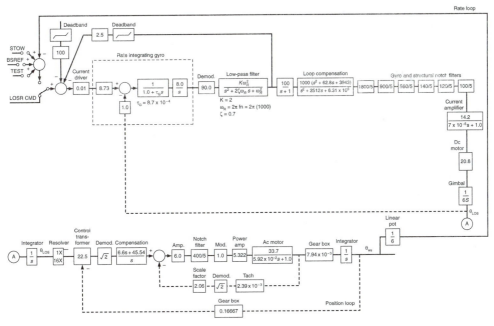

Figure 5.27 Line-of-sight stabilization pointing system elevation axis.

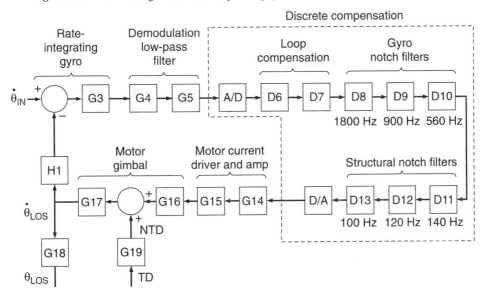

Figure 5.28 Line-of-sight stabilization/pointing system rate-loop line of sight.

Hardware. In the digital control system, the analog compensators and the notch filters are replaced by a digital signal processor, the TMS32010, along with the additional interface hardware needed to provide the digital input signals to the plant. Fig. 5.29 shows the system hardware block diagram.

The hardware was packaged onto five wire-wrap boards. It was fabricated as a prototype test bed and was constructed from commercially available components that have military-specification counterparts. Twelve-bit A/D converters were used, based on the studies of time-domain performance characteristics of the system.

Software. The TMS32010 software is composed of four modules: Initialization Routine, Main Program, Rate-Loop Subprogram, and Subroutines. Fig. 5.30 shows the system software block diagrams. The Initialization software disables and enables interrupts, loads data memory with filter coefficients, program constants, and gain terms, and initializes the TMS32010 registers.

The Main Program software calls the Delay Subroutine at the beginning of each sample period to wait for the A/D to complete conversion of all input variables. It then does on-line compensation for the error signal sensor variation by executing the A/D Drift Subroutine. The Main Program then reads the value of the input variable, calls the Rate-Loop Subprogram to compute the control output, and when that subprogram returns the output variable, loads it into the appropriate output register.

The Rate-Loop Subprogram calls subroutines that perform each compensator and notch filter computation and checks the computed output for flow. The subroutines consist of a single routine for performing any of the compensator or notch-filter computations (first- and second-order filter routine), along with routines for checking overflow, providing delay, and performing multiplication of low-precision numbers by a constant. The A/D Drift subroutine compensates on-line for the variation in rate-loop error signal sensor (as a function of time and temperature). The subroutine uses an external calibration input and follows the model of the sensor variations to estimate the true value of the A/D input.

The digital control system is interrupt driven. An interrupt occurs every 1/4020 second (approximately 250 µs). This starts the A/D conversion of a new set of sample inputs and restarts the TMS32010 on a new pass through its software.

The TMS32010 Evaluation Module (EVM) and Emulator (XDS) were used in the software development to permit single-step execution of the software for comparison with the corresponding computations produced by simulations written with the aid of the Continuous System Modeling Program (CSMP). These simulations take into account the input/output signal quantization levels, microprocessor architecture, memory, and internal register lengths. Other software functions associated with a complete, self-contained control module include:

1. System calibration, testing, and startup
2. Error checking and contingency responses
3. Setting of gains, time constants, and other programmable or adjustable parameters.
4. System shutdown

These functions are implemented by a general-purpose executive processor (SBP9989), thus allowing the TMS32010 to handle computation-intensive tasks.

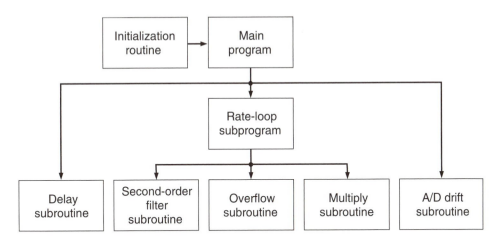

Figure 5.29 Digital controller hardware block diagram.

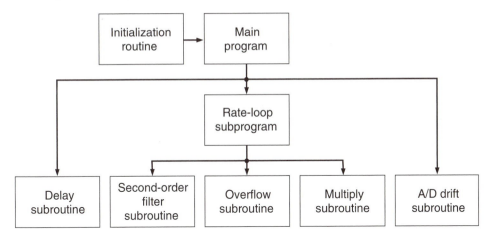

Figure 5.30 Digital controller software diagram.

System Performance. The system performance was evaluated in the following two-step procedure:

1. A hybrid computer system was constructed consisting of an analog computer implementation of part of the nondigital portion of the rate-loop coupled with the TMS32010-based digital controller.
2. A full-scale CSMP simulation of the entire rate loop was conducted.

The closed-loop performance of the rate loop was characterized by the following responses: rate-command step response, torque-disturbance step response, and torque-disturbance frequency response. The results are shown in Fig. 5.31 through 5.33.

In the rate-command step response, the percent overshoot, peak time, and settling time are similar for both the discrete and continuous systems, but the continuous system is slightly smoother. The torque-disturbance step response shows that the discrete system is slightly slower in correcting for a torque-disturbance input. In addition, the discrete system has a low-level oscillation (limit cycle). The frequency responses for a torque-disturbance input are also similar, with the continuous system having slightly better torque disturbance rejection in the low-frequency region. These results show that the analog and the digital systems are comparable even though no special efforts were made to take advantage of the capability that digital control offers.

The flexibility of the digital control system was demonstrated by programming the digital system with the capability to correct for a variation in the sensor input to the A/D converter. The system was able to compensate on-line (by using a known standard, calculating the gain, and dividing it out) for a 50 percent sinusoidal variation in the sensor gain.

The conversion between two different stabilization systems serves as another flexibility example. The software of a small, two-axis stabilization system was converted to the software of the higher-precision, large, two-axis gimbaled-platform stabilization loop described earlier. The only modification required was in the Main Program and the Rate-Loop Subprogram for the latter system. The modular software design procedure made possible the use of most of the building blocks (subroutines) in the implementation of the new controller.

In general, the study demonstrated the technical feasibility of digital control for a wide-bandwidth, high-precision type of system. Due to the limited scope of the study, the full power of digital control was not utilized, in that the control algorithms were constrained by the design to emulate their analog prototypes. It is likely that significant performance improvements could be achieved by advanced control techniques.

Additional capacity in the TMS32010 remains to accommodate improved, more sophisticated compensators. Table 5.10 shows the TMS32010 utilization.

Table 5.10 TMS32010 Utilization (LOS Stabilization System Rate-Loop)

Compensator/ Filter Element	Analog Transfer Function	Digital Transfer Function (f_s = 4020 Hz)
Rate-Loop First-order compensator	$G6 = \dfrac{100}{s+1}$	$D6 = \dfrac{0.1244(1.0 + 1.0z^{-1})}{1.0 - 0.99975z^{-1}}$
Second-order compensator	$G7 = \dfrac{1000(s^2 + 62.8s + 3943)}{s^2 + 2512s + 6.31 \times 10^6}$	$D7 = \dfrac{754.7101\,(1.0 - 1.98426z^{-1} + 0.9845z^{-2})}{1.0 - 1.255z^{-1} + 0.5474z^{-2}}$
1800-Hz notch filter	$G8 = \dfrac{s^2 + (2\pi \cdot 1800)^2}{s^2 + \frac{(2\pi \cdot 1800)}{5}s + (2\pi \cdot 1800)^2}$	$D8 = \dfrac{0.96877 + 1.83411z^{-1} + 0.9845z^{-2}}{1.0 + 1.83411z^{-1} + 0.93754z^{-2}}$
900-Hz notch filter	$G9 = \dfrac{s^2 + (2\pi \cdot 900)^2}{s^2 + \frac{(2\pi \cdot 900)}{2.5}s + (2\pi \cdot 900)^2}$	$D9 = \dfrac{0.8354 - 0.27291z^{-1} + 0.8352z^{-2}}{1.0 - 1.27291z^{-1} + 0.67041z^{-2}}$
560-Hz notch filter	$G10 = \dfrac{s^2 + (2\pi \cdot 560)^2}{s^2 + \frac{(2\pi \cdot 560)}{5}s + (2\pi \cdot 560)^2}$	$D10 = \dfrac{0.9287 - 1.19021z^{-1} + 0.9287z^{-2}}{1.0 - 1.19021z^{-1} + 0.8574z^{-2}}$
140-Hz notch filter	$G11 = \dfrac{s^2 + (2\pi \cdot 140)^2}{s^2 + \frac{(2\pi \cdot 140)}{5}s + (2\pi \cdot 140)^2}$	$D11 = \dfrac{0.97875 - 1.91083z^{-1} + 0.97875z^{-2}}{1.0 - 1.91083z^{-1} + 0.95751z^{-2}}$
120-Hz notch filter	$G12 = \dfrac{s^2 + (2\pi \cdot 120)^2}{s^2 + \frac{(2\pi \cdot 120)}{5}s + (2\pi \cdot 120)^2}$	$D12 = \dfrac{0.9817 - 1.92896z^{-1} + 0.9817z^{-2}}{1.0 - 1.92896z^{-1} + 0.96339z^{-2}}$
100-Hz notch filter	$G13 = \dfrac{s^2 + 62.8S + 3943)}{s^2 + \frac{(2\pi \cdot 100)}{5}s + (2\pi \cdot 100)^2}$	$D13 = \dfrac{0.98467 - 1.94534z^{-1} + 0.98467z^{-2}}{1.0 - 1.94534z^{-1} + 0.96935z^{-2}}$
Position-Loop First-order compensator	$G22 = \dfrac{6.6s + 45.54}{s}$	$D22 = \dfrac{6.60566 - 6.59434z^{-1}}{1.0 - z^{-1}}$
400-Hz notch filter	$G24 = \dfrac{s^2 + (2\pi \cdot 400)^2}{s^2 + \frac{(2\pi \cdot 400)}{5}s + (2\pi \cdot 400)^2}$	$D24 = \dfrac{0.94471 - 1.53204z^{-1} + 0.94471z^{-2}}{1.0 - 1.53204z^{-1} + 0.94471z^{-2}}$

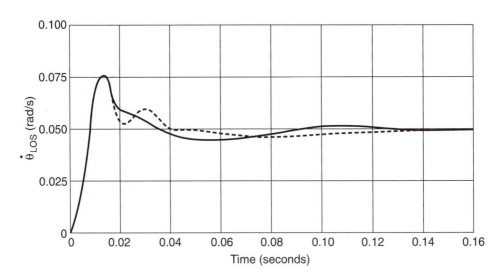

Figure 5.31   Rate-loop rate command step response.

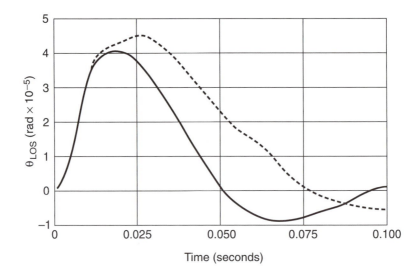

Figure 5.32   Rate loop normalized torque disturbance step response.

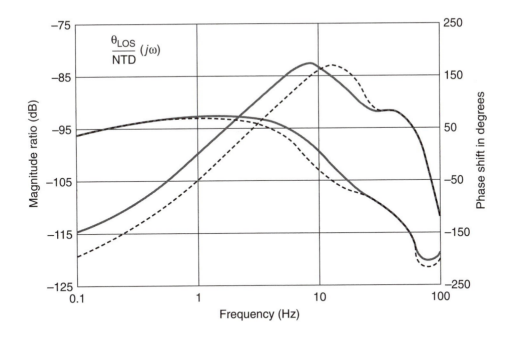

Figure 5.33   Rate-loop normalized torque disturbance frequency response.

Other microprocessors that were considered for implementing this digital controller system were: Intel 8086, Zilog Z-8000, Motorola 68000, and the Fairchild 9445. These microprocessors were unable to meet the criterion that the maximum allowable time between samples for processing be 250 μS. Among the signal-processing microprocessors, the AMI 2811, while apparently fast enough, has only a 12 x 12 multiplier, and the Intel 2920 has only four inputs and no branching instructions.

The principal limitation of the TMS32010 was that of having eight inputs and eight outputs. Except for this restriction, the processor would have been able to carry out the processing for both axes of the two-axis gimbaled platform. This limitation could be removed by the addition of logic circuitry.

5.5 SUMMARY

Table 5.1 summarizes the differences between analog and digital signal processing techniques. Their programmability makes DSPs an attractive alternative to conventional signal processors. Algorithms, sample rates, and clock rates are key characteristics of DSP systems. In control systems, the sampling rate selection is usually about 10 to 20 times the system's bandwidth.

Table 5.2 compares a single-chip DSP with a general-purpose microprocessor. Because of their special architecture, DSPs are used for implementation of advanced mathematical algorithms. Table 5.3 compares fixed- and floating-point arithmetic types of DSPs. Architectural features of DSPs and their benefits are summarized in Table 5.4. Typical benchmarked operations for DSPs are matrix multiplication, complex multiplication, and PID loop calculations. Other DSP solutions include DSP cores, DSP multiprocessors, and chip sets.

In Section 5.4 we discuss how commercially, available DSPs are used in specific control system applications. Motor control, robotics, automotive control, stabilization control, and disk-drive control systems were considered. In Chapter 6, we cover more DSP applications in control systems. These applications are based on modern control system techniques such as fuzzy logic, observability, and controllability methods.

PROBLEM

5.1 Consider the system in Fig. 5.34. For the closed-loop yaw-axis control system present in this block diagram with the simplified aircraft dynamics represented by the transfer function

$G_s(s)$:

(a) Determine a PID controller with unity dc gain that yields a 45-degree Phase Margin. (The PID design material from Chapter 4 will be useful for this problem).
(b) Implement this controller on two different DSPs (e.g., TI and Analog Devices.)
(c) Compare the two DSP solutions.

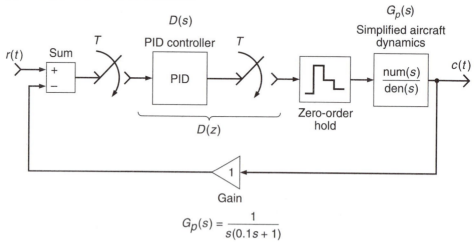

$$G_p(s) = \frac{1}{s(0.1s + 1)}$$

Figure 5.34   Yaw-axis attitude control system block diagram.

REFERENCES

1. *ADSP-2101/2102 User's Manual,* Analog Devices, Inc., Norwood, MA, February 1990.

2. *Digital Signal Processing Applications with the TMS320 Family, Theory: Algorithms, and Implementations,* Application Book Vol. 1, Texas Instruments, Dallas, TX, 1989.

3. *Digital Signal Processing Applications with the TMS320 Family: Theory, Algorithms, and Implementations,* Application Book Vol. 2, Texas Instruments, Dallas, TX, 1990.

4. *Digital Signal Processing Applications with the TMS320 Family: Theory, Algorithms, and Implementations,* Application Book Vol. 3, Texas Instruments, Dallas, TX, 1990.

5. *Digital Signal Processing Applications with the TMS320 Family*: *Selected Applications Notes,* Application Book, Texas Instruments, Dallas, TX, 1991.

6. *TMS320 Family Development Support Reference Guide,* Texas Instruments, Dallas, TX, 1990.

7. *TMS320 Third-Party Support Reference Guide,* Texas Instruments, Dallas, TX, 1990.

8. *TMS320C1X User's Guide,* Texas Instruments, Dallas, TX, 1991.

9. *Third-Generation TMS320 User's Guide,* Digital Signal Processor Products, Texas Instruments, Dallas, TX, 1988.

10. *TMS320 User's Guide,* Digital Signal Processor Products, Texas Instruments, Dallas, TX, 1986.

11. *Digital Signal Processing Applications with the TMS320 Family: Theory, Algorithms, and Implementations*, Texas Instruments, Dallas, TX, 1986.

12. *Digital Control Applications with the TMS320 Family: Selected Application Notes,* Texas Instruments Applications Book, Dallas, TX, 1991.

13. *DSP 56000/DSP 56001 Digital Signal Processor User's Manual,* Motorola Inc., 1990.

14. *WE DSP 32 Digital Signal Processor Information Manual,* AT&T Corporation, 1988.

15. Analog Devices, Inc., *Digital Signal Processing Applications*: *Using the ADSP-2100 Family,* Prentice Hall, Upper Saddle River, NJ, 1990.

16. R. Chassaing and D.W. Horning, *Digital Signal Processing with the TMS320C25,* Wiley, New York, 1990.

17. A.V. Oppenheim and R. V. Schafer, *Discrete-Time Signal Processing,* Prentice Hall, Upper Saddle River, NJ, 1989.

18. *Digital Signal Processing, Single-Chip DSP Processors, Theory-Design and Applications,* DSP Associates, 1989.

19. *Control Theory, Implementation, and Applications with the TMS320* Family Seminar Workbook, Texas Instruments, Dallas, TX, 1991.

20. *Considerations for Selecting a DSP Processor (ADSP-2100 A vs. TMS320C25),* Application Note, Analog Devices, Inc., Norwood, MA, June 1989.

21. *Considerations for Selecting a DSP Processor (ADSP-2100 A vs. WEDSP16A),* Application Note, Analog Devices, Inc., Norwood, MA, August 1990.

22. *Disk-Drive Servo Control with WEDSP16/DSP16A Digital Signal Processor,* Application Note, AT&T Corporation, June 1989.

23. *Implementation of PID Controllers on the Motorola DSP56000/DSP56001,* Motorola, Inc., 1989.

24. *TMS320C24x DSP Controller,* Product Bulletin, Texas Instruments, Dallas, TX, 1996.

25. *Details on Signal Processing,* Texas Instruments, Dallas, TX, September 1995.

26. P. Lapsley, J. Bier, A. Shoham, and E. Lee, *DSP Processor Fundamentals, Architectures and Features,* Berkeley Design Technology, Inc., 1996.

    Berkeley Design Technology, Inc. (BDTI)
    2107 Dwight Way, Second Floor
    Berkeley, CA 94707
    (510) 665-1600: TEL
    (510) 665-1680: FAX
    info@bdti.com : email
    http://www.bdti.com

27. *AC Motor Control Using the ADMC200 Coprocessor,* Application Note, Analog Devices, Inc., Norwood, MA.

28. T. V. Raumer, J. M. Dion, L. Dugard, and J. L. Thomas, Applied Nonlinear Control of an Induction Motor Using Digital Signal Processing, *IEEE Transactions on Control Systems Technology,* vol. 2, no. 4, December 1994.

29. *Digital Signal Processing Products,* Texas Instruments, Dallas, TX, 1994.

30. *dSPACE Product Information,* dSPACE GmbH, Paderborn, Germany, 1990.

31. C. Slivinsky and J. Borninski, *Control System Compensation and Implementation with the TMS32010 Digital Signal Processing Applications with the TMS320 Family: Theory, Algorithms, and Implementations; Application Book Vol. 1,* Texas Instruments, Dallas, TX, 1989.

32. H. Hanselman, "Implementation of Digital Controllers—A Survey", *Automatica,* Pergamon Press, vol. 23, no.1, 1987.

33. Galil, *Motion Control Product Catalog, 1995.*

CHAPTER **6**

Modern Design Techniques and Their Applications

6.1 INTRODUCTION

In Chapter 5 we investigated the use of DSPs to implement controllers, which we represented in transfer function form as $D(z)$ or $G_c(z)$. These transfer functions, which need to be written in difference equation form to be implemented with DSPs, had been arrived at via classical design procedures, such as the root-locus and frequency-domain techniques. Rather than being based on representations of physical systems in transfer function form, modern design techniques for digital control system design, which is the focus of this chapter, are generally based on the state-variable model of a system. The state-space formulations presented in Chapter 3 are employed extensively in this chapter. As we mentioned in Chapter 3, these methods are useful in MIMO (multiple-input/multiple-output) systems, whereas the classical transfer function approach is primarily a SISO (single-input/single-output) methodology. In Chapter 3 we also mentioned that there are CAD tools available for transforming the transfer function description of a system into a state-space formulation, and vice versa (see Appendix A). These, as well as other CAE/CAD tools, are important in this chapter.

Once a controller description is determined from design techniques, whether classical or modern, that controller can be implemented with DSP technology. As long as our systems are assumed to be linear, modern-state variable design techniques will yield a linear controller that can be cast into a $D(z)$ or $G_c(z)$ form, or into a matrix of such transfer functions in the MIMO system case. From that point on, the designer just needs to implement the transfer function with a DSP. Whether the controller transfer function came from classical frequency-domain design procedures or from state-variable methods makes little difference at the implementation stage of the game. The performance of the system is the essential matter. Typically, state-variable methods will yield a better performance, but at the cost of more complex mathematics and transfer functions. After all, in most classical control problems we have only the plant output to use as feedback. In state-variable design, on the other hand, we have potentially all the system states to serve as feedback. Intuitively, that should give us a better design. How sophisticated we make our controllers, as usual, depends on how much we are willing to spend on the design.

Section 6.2 covers the design of discrete systems via the pole-placement procedure. Controllability is assumed for this procedure. We investigate the definition of and various tests for controllability. The controllability canonical form (CF) introduced in Chapter 3 will be invoked.

Section 6.3 covers the business of observers and observability. Observers or estimators provide the designer with estimates of the state variables, which may be difficult or impossible to determine by direct measurement. But to construct an observer it is necessary that the system be observable. The definition of and tests for observability are considered. The observability canonical form (OF) will be invoked from Chapter 3.

The basic ideas associated with linear quadratic optimal (LQ) design are taken up in Section 6.4. Since that topic is vast, only a few results are presented. The MATLAB and MATRIX$_X$ software packages both have procedures to solve the LQ problem rapidly. Examples using these tools are presented.

Section 6.5 considers a very recent control methodology that is growing in popularity: fuzzy logic control (FLC). The literature on fuzzy logic indicates that it is an excellent procedure to employ when the plant model is unknown or has nonlinearities that defy simple analysis. We touch only on the basic ideas associated with the control of systems using fuzzy logic, but will present an example that generates an FLC design and then implements that design on a DSP.

6.2 CONTROLLABILITY AND POLE-PLACEMENT

Controllability

To introduce the notion of *controllability* consider the system presented in Fig. 6.1. Note that the state $x_2(k)$ cannot be affected by the system input $r(k)$. That means that the system cannot be fully controlled, even though the other state $x = Tx_c$ can be accessed.

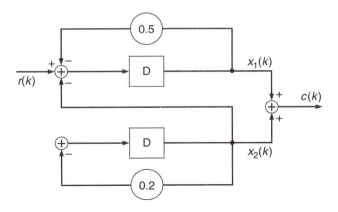

Figure 6.1   Uncontrollable system.

Definition 6.1 If there exists an input $r(k)$ that transfers any given initial state vector $x_0 = x(0)$ of the system

$$x(k + 1) = Ax(k) + Br(k) \qquad (6.1)$$

to a desired state $x(N)$ in a finite number of sampling instances, that system is said to be completely state controllable or, more simply, just controllable.

The system presented in Fig. 6.1 is obviously uncontrollable, but many systems, even if their block diagram appears controllable, can in fact be uncontrollable. Consider the system in Fig. 6.2. It looks controllable but actually is not.

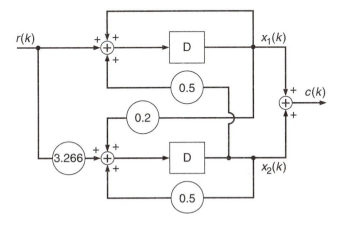

Figure 6.2   Another uncontrollable discrete-time system.

Fortunately, several tests have been developed to determine analytically whether a given system is or is not controllable. We present only one test, which is based on the notion of a controllability matrix. It is given here without proof. A proof is given in Chapter 9 of Reference 1.

Theorem 6.1 A discrete-time system is controllable if and only if the controllability matrix

$$V = \left[B|AB|A^2B|B\ldots A^{n-1}B \right]_{nx(np)} \qquad (6.2)$$

has rank n.

Note that only the state equation (A, B) and not the output equation (C, D) is involved in this determination. This result holds for SISO and MIMO systems.

As mentioned in Appendix E, the *rank* of a matrix V, call it r (V), is the number of linearly independent rows or columns of V. But it is not always easy to determine whether rows or columns are or are not linearly dependent. An indirect way to make this determination, however, is to determine the largest nonsingular submatrix V contains. See if determinants of these submatrices are zero. Then r(V) equals the order of the largest nonsingular submatrix that V contains.

Example 6.1

Consider a discrete-time system and its state equation

$$x(k + 1) = \begin{bmatrix} 0 & 1 \\ -2 & 3 \end{bmatrix} x(k) + \begin{bmatrix} 1 \\ 1 \end{bmatrix} r(k)$$

Because n = 2, from Equation 6.2,

$$V = [B \ AB] = \begin{bmatrix} 1 & 1 \\ 1 & 1 \end{bmatrix}$$

and the rank of the matrix is 1. Therefore, the discrete system is not controllable.

Example 6.2

Consider the MIMO system

$$A = \begin{bmatrix} 0.2 & 1 \\ -1 & 0.5 \end{bmatrix}$$

$$B = \begin{bmatrix} 0 & 1 \\ 1 & 0.2 \end{bmatrix}$$

Form the controllability matrix

$$V = [B \ AB] = \begin{bmatrix} 0 & 1 & 1 & 0.4 \\ 1 & 0.2 & 0.5 & -0.9 \end{bmatrix}$$

which has several nonsingular 2 x 2 submatrices. Therefore, the system is controllable.

Controllability is actually a property of a system representation, not of a system per se. We would like to find a transformation technique that transforms any discrete-time system into one that is controllable. Consider the system of Fig. 6.3. The figure shows transformation of a system into a controllable form according to the mapping rule $x = Tx_c$

where $A_c = \begin{bmatrix} 0 & 1 \\ -0.1 & 0.7 \end{bmatrix}$ is the controllable form of the state x and T is the

transformation matrix.

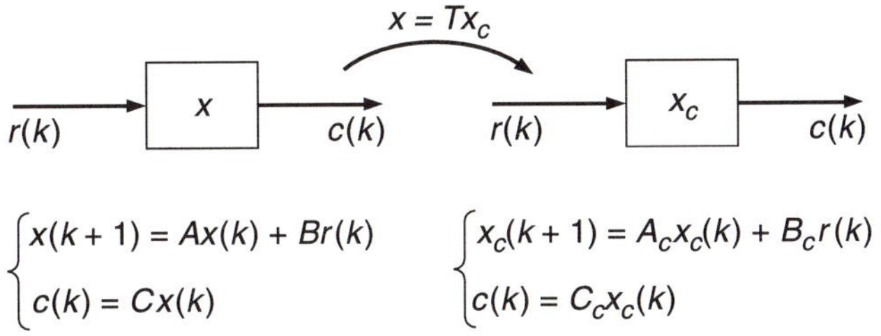

$$\begin{cases} x(k+1) = Ax(k) + Br(k) \\ c(k) = Cx(k) \end{cases} \qquad \begin{cases} x_c(k+1) = A_c x_c(k) + B_c r(k) \\ c(k) = C_c x_c(k) \end{cases}$$

Figure 6.3   Transformation of a system into a controllable form.

The controllable dynamical equations consist of the matrices A_c, B_c, and C_c, where $A_c = T^{-1}AT$, $B_c = T^{-1}B$, and $C_c = CT$. As discussed in Chapter 3, this is a canonical form, the controllability canonical form. Matrices A_c, B_c, and C_c can be determined by reading coefficients from the system transfer function. The method of Chapter 3 required determination of the system transfer function, given an arbitrary system representation A, B, and C, for which we can use Equation 3.82. (Assume that for simplicity the D matrix is zero.) In this chapter we present a method for generating the CF without needing explicitly to determine the system transfer function. What we do need is the characteristic equation

$$|\lambda I - A| = \lambda^n + a_{n-1}\lambda^{n-1} + \cdots + a_1\lambda + a_0 = 0 \tag{6.3}$$

and matrices A_c and B_c have the following forms:

$$A_c = \begin{bmatrix} 0 & 1 & 0 & \cdots & 0 & 0 \\ 0 & 0 & 1 & \cdots & 0 & 0 \\ & & \vdots & & & \\ 0 & 0 & 0 & \cdots & 0 & 1 \\ -a_0 & -a_1 & -a_2 & \cdots & -a_{n-2} & -a_{n-1} \end{bmatrix} \tag{6.4}$$

$$B_c = \begin{bmatrix} 0 \\ 0 \\ \vdots \\ 1 \end{bmatrix} \tag{6.5}$$

where the bottom row of A consists of the coefficients of the characteristic equation. Once we know matrices A_c and B_c, we can find the corresponding controllability matrix V_c:

$$V_c = \left[B_c \vdots A_c B_c \vdots \cdots \vdots A_c^{n-1} B_c \right]. \tag{6.6}$$

The question remains: How do we determine C_c? In Chapter 3 it was available from the numerator of the transfer function. But here we want to avoid the transfer function calculation. One approach is to look at the controllability matrices. From Equation 6.6 we have

$$\begin{aligned} V_c &= \left[B_c \vdots A_c B_c \vdots \cdots \right] \\ &= \left[T^{-1}B \vdots T^{-1}ATT^{-1}B \vdots \cdots \right] \\ &= T^{-1}\left[B \vdots AB \vdots \cdots \right] \\ &= T^{-1}V \end{aligned} \tag{6.7}$$

or

$$V = TV_c \qquad (6.8)$$

which can be solved for T. In fact, for SISO systems the controllability matrices are square and V_c is nonsingular. Thus we can write the transformation matrix T as

$$T = VV_c^{-1} \qquad (6.9)$$

and

$$C_c = CT \qquad (6.10)$$

Example 6.3

Cast the system

$$A = \begin{bmatrix} 0.5 & 0 \\ 1 & 0.2 \end{bmatrix}$$

$$B = \begin{bmatrix} -1 \\ 1 \end{bmatrix}$$

$$C = \begin{bmatrix} 2 & 0.5 \end{bmatrix}$$

into the controllability canonical form.

Solution: First we form the characteristic equation

$$|\lambda I - A| = 0 = \det \begin{bmatrix} \lambda - 0.5 & 0 \\ -1 & \lambda - 0.2 \end{bmatrix} = \lambda^2 - 0.7\lambda + 0.1$$

from which we construct

$$A_c = \begin{bmatrix} 0 & 1 \\ -0.1 & 0.7 \end{bmatrix}$$

and assuming that $B_c = \begin{bmatrix} 0 \\ 1 \end{bmatrix}$ we need only determine C_c. The controllability matrices are

$$V = [B \vdots AB] = \begin{bmatrix} -1 & -0.5 \\ 1 & -0.8 \end{bmatrix}$$

and

$$V_c = [B_c \vdots A_c B_c] = \begin{bmatrix} 0 & 1 \\ 1 & 0.7 \end{bmatrix}$$

Therefore,

$$T = VV_c^{-1} = \begin{bmatrix} 0.2 & -1 \\ -1.5 & 1 \end{bmatrix}$$

and

$$C_c = CT = \begin{bmatrix} 2 & 0.5 \end{bmatrix} \begin{bmatrix} 0.2 & -1 \\ -1.5 & 1 \end{bmatrix} = \begin{bmatrix} -0.35 & -1.5 \end{bmatrix}$$

As a check on this method, we use the method from Chapter 3, for which we need the transfer function

$$H(z) = C(zI - A)^{-1}B = \begin{bmatrix} 2 & 0.5 \end{bmatrix} \begin{bmatrix} z - 0.5 & 0 \\ -1 & z - 0.2 \end{bmatrix}^{-1} \begin{bmatrix} -1 \\ 1 \end{bmatrix}$$

$$= \begin{bmatrix} 2 & 0.5 \end{bmatrix} \begin{bmatrix} z - 0.2 & 0 \\ -1 & z - 0.5 \end{bmatrix} \frac{1}{\Delta} \begin{bmatrix} -1 \\ 1 \end{bmatrix}$$

where $\Delta = (z - 0.2)(z - 0.5)$. Carrying out the multiplications, we get

$$H(z) = \frac{-1.5z - 0.35}{z^2 - 0.7z + 0.1}$$

and picking off the coefficients, we get, as we would expect,

$$A_c = \begin{bmatrix} 0 & 1 \\ -0.1 & 0.7 \end{bmatrix}$$

$$B_c = \begin{bmatrix} 0 \\ 1 \end{bmatrix}$$

$$C_c = \begin{bmatrix} -0.35 & -1.5 \end{bmatrix}$$

Example 6.4

Consider the third-order system representation

$$A = \begin{bmatrix} 1 & 0 & 0 \\ 0 & 0.5 & 0 \\ 0 & 0 & 0.25 \end{bmatrix}$$

$$B = \begin{bmatrix} 1 \\ 1 \\ 0 \end{bmatrix}$$

$$C = \begin{bmatrix} 1 & -1 & 1 \end{bmatrix}$$

which can be seen to be uncontrollable because no control effect can be brought to bear on the third state in the state vector. Checking the controllability matrix, we get

$$V = \begin{bmatrix} B & AB & A^2B \end{bmatrix} = \begin{bmatrix} 1 & 1 & 1 \\ 1 & 0.5 & 0.25 \\ 0 & 0 & 0 \end{bmatrix}$$

which has rank 2 instead of rank 3, which would be required for controllability. Now, transforming the system to the CF, we need the characteristic equation

$$|\lambda I - A| = 0 = \det \begin{bmatrix} \lambda - 1 & 0 & 0 \\ 0 & \lambda - 0.5 & 0 \\ 0 & 0 & \lambda - 0.25 \end{bmatrix} = \lambda^3 - 1.75\lambda^2 + 0.875\lambda - 0.125$$

from which we form

$$A_c = \begin{bmatrix} 0 & 1 & 0 \\ 0 & 0 & 1 \\ 0.125 & -0.875 & 1.75 \end{bmatrix} \text{ and knowing that } B_c = \begin{bmatrix} 0 \\ 0 \\ 1 \end{bmatrix}$$

all that remains is to determine C_c. We need

$$V_c = \begin{bmatrix} B_c & A_c B_c & A_c^2 B_c \end{bmatrix} = \begin{bmatrix} 0 & 0 & 1 \\ 0 & 1 & 1.75 \\ 1 & 1.75 & 2.1875 \end{bmatrix}$$

which has full rank, indicating a controllable system representation. Then

$$T = VV_c^{-1} = \begin{bmatrix} 1 & 1 & 1 \\ 1 & 0.5 & 0.25 \\ 0 & 0 & 0 \end{bmatrix} \begin{bmatrix} 0 & 0 & 1 \\ 0 & 1 & 1.75 \\ 1 & 1.75 & 2.1875 \end{bmatrix}^{-1} = \begin{bmatrix} 0.125 & -0.75 & 1 \\ 0.25 & -1.25 & 1 \\ 0 & 0 & 0 \end{bmatrix}$$

and

$$C_c = CT = \begin{bmatrix} 1 & -1 & 1 \end{bmatrix} \begin{bmatrix} 0.125 & -0.75 & 1 \\ 0.25 & -1.25 & 1 \\ 0 & 0 & 0 \end{bmatrix} = \begin{bmatrix} -0.125 & 0.5 & 0 \end{bmatrix}$$

The reader is invited to compute the transfer function and determine the CF via the Chapter 3 method.

Once a controllable system representation is arrived at, the next step is to control the system in some way or other. A simple but elegant approach is to control the system by placing its closed-loop poles in desirable locations of the z-plane. This pole placement procedure can be accomplished with state variable feedback.

Pole-Placement

The method of pole-placement requires that we know the state vector of the system, and in fact that the control variable is constituted by an appropriate linear combination of states. Measurement of all the system states (velocity, current, etc.) is normally impractical and sometimes impossible. We must usually estimate the state vector based on the available system inputs and outputs. But we assume initially that all the states are available, after which we investigate the more realistic situation wherein the states need to be estimated. Consider the system of Fig. 6.4, with its corresponding dynamical equations

$$x(k+1) = Ax(k) + Br(k) \qquad\qquad (6.11)$$

$$c(k) = Cx(k) + Dr(k) \qquad\qquad (6.12)$$

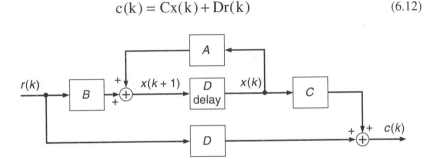

Figure 6.4   Discrete-time system.

The placement of the poles of the closed-loop system characteristic equation at their desired locations requires the following steps:

1. Define the input r(k) by

$$r(k) = -Kx(k) + \hat{r}(k) \qquad\qquad (6.13)$$

where $K = \begin{bmatrix} k_1 & k_2 & \cdots & k_n \end{bmatrix}$ is the gain matrix and $\hat{r}(k)$ is a reference input. A special case of Equation 6.13, normally called the *control law*, applies to the design of regulators if the reference input is zero. In this type of system, the feedback element consists of a linear combination of all states, and we assume for simplicity that the outputs are the states, i.e., $c(k) = x(k)$. Fig. 6.5 shows a control law in a discrete time control system regulator.

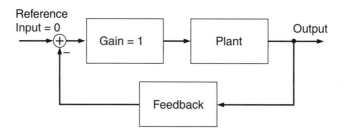

Figure 6.5   Role of a control law in a discrete-time control system regulator.

In the more general case, substituting our control law equation into the original state equation, we obtain

$$x(k+1) = (A - BK)x(k) + B\hat{r}(k)$$ (6.14)

2. Choose the desired pole locations $z = p_1, p_2 \dots p_n$. The discrete characteristic equation of the system is

$$\det[zI - A + BK] = 0$$ (6.15)

or

$$\alpha_d(z) = (z - p_1)(z - p_2) \cdots (z - p_n)$$ (6.16)

3. Find the elements of the gain matrix K so that the roots of the characteristic equation are at the desired locations. This procedure is reduced to equating Equations 6.15 and 6.16 and solving for elements of K.

Example 6.5

Consider the system of Example 6.4, which we transformed into

$$A_c = \begin{bmatrix} 0 & 1 & 0 \\ 0 & 0 & 1 \\ 0.125 & -0.875 & 1.75 \end{bmatrix} \text{ and } B_c = \begin{bmatrix} 0 \\ 0 \\ 1 \end{bmatrix}$$

The eigenvalues or poles of this system are at z = -1.0, -0.5, and -0.25. Let us assume that we want all three poles located at z = 0.5. We can form the equation

$$\det[zI\text{-}A\text{+}BK] = \alpha_d(z) = (z - p_1)(z - p_2)\cdots(z - p_n)$$

$$\det\left\{\begin{bmatrix} z & 0 & 0 \\ 0 & z & 0 \\ 0 & 0 & z \end{bmatrix} - \begin{bmatrix} 0 & 1 & 0 \\ 0 & 0 & 1 \\ 0.125 & -0.875 & 1.75 \end{bmatrix} + \begin{bmatrix} 0 \\ 0 \\ 1 \end{bmatrix}\begin{bmatrix} k_1 & k_2 & k_3 \end{bmatrix}\right\} = (z - 0.5)^3$$

$$z^3 + z^2(k_3 - 1.75) + z(k_2 + 0.875) + k_1 - 0.125 = z^3 - 1.5z^2 + 0.75z - 0.125$$

and equating coefficients, we get

$$k_3 - 1.75 = -1.5$$
$$k_2 + 0.875 = 0.75$$
$$k_1 - 0.125 = -0.125$$

from which it follows that

$$(F_s = \frac{1}{T_s})$$
$$k_2 = -0.125$$
$$k_3 = 0.25$$

These equations were very easy to solve because the system was in the controllability canonical form. The solution for the k terms involves only uncoupled algebraic equations. If other forms of the A matrix are involved, the solution for the k terms generally involves solving n equations in n unknowns. Fortunately, there are several CAE tools available to compute the gains in the state-variable feedback problem. Most of these take advantage of the mathematical properties of systems represented in the controllability canonical form and are based on a procedure, discussed next, due to J. Ackermann.

Ackermann's Formula

The procedure explained above to compute the gain K can in general be a very tedious task. A more convenient way of calculating the gain matrix K is via a procedure called *Ackermann's formula*. Consider the desired characteristic equation

$$\alpha_{d}(z) = (z - p_{1})(z - p_{2}) \cdots (z - p_{n})$$ (6.17)

Since the Cayley-Hamilton theorem says that every 'A" matrix satisfies its characteristic equation, this equation should be zero if we substitute the closed-loop A matrix in place of z. Let that closed-loop A matrix be $\hat{A}$ where

$$\hat{A} = A - BK$$ (6.18)

In other words, from the Cayley-Hamilton theorem,

$$\alpha_{d}(\hat{A}) = 0$$ (6.19)

But for the Ackermann formula we need

$$\alpha_{d}(A) = A^{n} + \alpha_{n-1}A^{n-1} + \cdots + \alpha_{1}A + \alpha_{0}I$$ (6.20)

which is generally not equal to zero and where A is the open-loop system matrix of the state equation. Then the formula derived by Ackermann for the feedback gain matrix K is

$$K = \begin{bmatrix} 0 & 0 & \cdots & 1 \end{bmatrix} \begin{bmatrix} B & AB & \cdots & A^{n-1}B \end{bmatrix}^{-1} \alpha_{d}(A)$$ (6.21)

where n is the order of the system. The middle term is recognizable as the controllability matrix V, so we can write

$$K = \begin{bmatrix} 0 & 0 & \cdots & 1 \end{bmatrix} V^{-1} \alpha_{d}(A).$$ (6.22)

Example 6.6

Consider the discrete system of Example 6.5:

$$A = \begin{bmatrix} 0 & 1 & 0 \\ 0 & 0 & 1 \\ 0.125 & -0.875 & 1.75 \end{bmatrix} \text{ and } V = \begin{bmatrix} 0 & 0 & 1 \\ 0 & 1 & 1.75 \\ 1 & 1.75 & 2.1875 \end{bmatrix}$$

$$\alpha_{d}(A) = A^{n} + \alpha_{n-1}A^{n-1} + \cdots + \alpha_{1}A + \alpha_{0}I = (A - 0.5I)^{3}$$

The matrix inverse and multiplications are performed with the MATRIX$_X$ software package and we get K = [0 -0.125 0.25] exactly as in Example 6.5.

Now let us consider the more realistic problem of what to do if the state variables are not available for direct measurement. In the case of a fifth-order system, for instance, we would generally need to have all five states available to implement a pole-placement design. That could be very expensive just in terms of sensors. We discuss the design of estimators to provide state approximations, but to do this we need to assume that our systems are observable.

6.3 OBSERVABILITY AND STATE ESTIMATION

Observability

To introduce the notion of *observability* consider the system represented in Fig. 6.6. Note that without knowledge of initial conditions the state $x_2(k)$ cannot be determined or observed by knowing the input and measuring the output. We cannot fully estimate the state vector by only a knowledge of r(k) and c(k). The system is said to be *unobservable*.

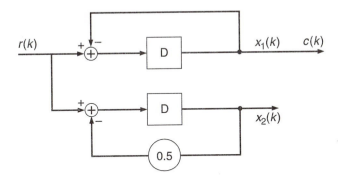

Figure 6.6   Unobservable system.

Definition 6.2   A discrete-time system

$$x(k+1) = Ax(k) + Br(k) \qquad (6.23)$$

$$c(k) = Cx(k) + Dr(k) \qquad (6.24)$$

is said to be *completely state observable* or, more simply, just *observable* if we can find $x(0)$ from the output c(k) and input r(k) sequences over a finite number of steps or measurements. By implication, since

$$x(k) = A^k x(0) + \sum_{n=0}^{k-1} A^{k-n-1} Br(k)$$

(6.25)

once we have $x(0)$ we have the state vector for all k.

As in controllability, the companion concept to observability, we may be able to determine the observability of a given system by inspection of a block diagram. But again, such determination may not be so obvious if the system is high order or complex. Fortunately, analytical tests are available to make the observability determination. We present one such test in the following theorem.

Theorem 6.2 The system represented by Equations 6.23 and 6.24 is observable if and only if its observability matrix defined by Equation 6.26 has rank n:

$$O = \begin{bmatrix} C \\ CA \\ CA^2 \\ \vdots \\ CA^{n-1} \end{bmatrix}$$

(6.26)

where n is the order of the system.

Example 6.7

The following dynamical equations describe a discrete-time system:

$$x(k + 1) = \begin{bmatrix} 0 & 1 \\ 0 & 2 \end{bmatrix} x(k) + \begin{bmatrix} 0 \\ 3 \end{bmatrix} r(k)$$

$$c(k) = \begin{bmatrix} 1 & 1 \end{bmatrix} x(k)$$

Because n = 2, Equation 6.26, the observability matrix is

$$O = \begin{bmatrix} C \\ CA \end{bmatrix} = \begin{bmatrix} 1 & 1 \\ 0 & 3 \end{bmatrix}$$

The rank of O is 2, and the system is observable.

Example 6.8

The following dynamical equations describe a MIMO discrete-time system:

$$x(k+1) = \begin{bmatrix} 0 & -1 \\ -1 & 0 \end{bmatrix} x(k) + \begin{bmatrix} 1 & 2 \\ -1 & 1 \end{bmatrix} r(k)$$

$$c(k) = \begin{bmatrix} 1 & -1 \\ -1 & 1 \end{bmatrix} x(k)$$

Again, because n = 2, the observability matrix is

$$O = \begin{bmatrix} C \\ CA \end{bmatrix}$$

but now because the C matrix is 2 x 2 we get

$$O = \begin{bmatrix} 1 & -1 \\ -1 & 1 \\ 1 & -1 \\ -1 & 1 \end{bmatrix}$$

and there are no 2 x 2 submatrices of this that have nonzero determinants. Therefore, the system is unobservable.

Observability, like controllability, is a property of a system representation, not of a system per se. And there is a system representation, the observability canonical form discussed in Chapter 3, which is always observable. We can define a mapping scheme in which a representation with state x is transformed to an observable form with state vector x_o. Similar to Fig. 6.3, where we defined the transformation of a system into a controllable form, Fig. 6.7 shows transformation of a system into an observable form.

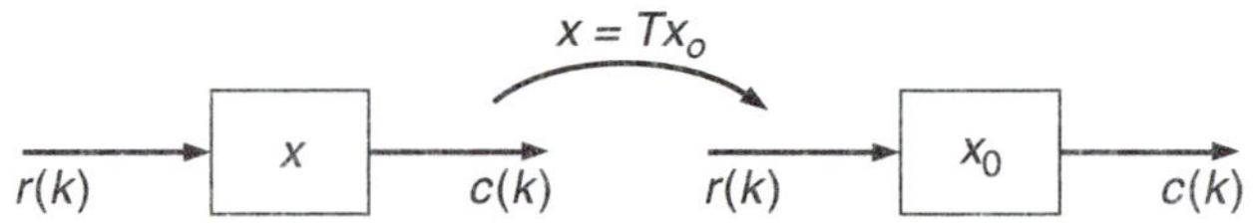

$$x = Tx_0$$

$$r(k) \rightarrow \boxed{x} \rightarrow c(k) \qquad r(k) \rightarrow \boxed{x_0} \rightarrow c(k)$$

$$\begin{cases} x(k+1) = Ax(k) + Br(k) \\ c(k) = Cx(k) \end{cases} \qquad \begin{cases} x_0(k+1) = A_o X_c(k) + B_o r(k) \\ c(k) = C_o x_c(k) \end{cases}$$

Figure 6.7   Transformation of a system into an observable form.

Let the mapping rule between given state vector and observable state vector be $x = Tx_0$. The observable dynamical equations consist of the matrices A_o, B_o, C_o where $A_o = T^{-1}AT$, $B_o = T^{-1}B$, and $C_o = TC$. As in Chapter 3, the matrices A_o, B_o, C_o can be determined by reading coefficients from the system transfer function. In this chapter we present a method that does not require transfer function determination but requires just the characteristic equation, Equation 6.3, from which we read the coefficients. The coefficient matrices A_o and C_o are

$$A_o = A_c^{\,T} = \begin{bmatrix} 0 & 0 & \cdots & 0 & -a_0 \\ 1 & 0 & \cdots & 0 & -a_1 \\ & & \vdots & & \\ 0 & 0 & \cdots & 1 & -a_{n-1} \end{bmatrix} \tag{6.27}$$

$$C_o = B_c^{\,T} = \begin{bmatrix} 0 & 0 & \cdots & 1 \end{bmatrix}. \tag{6.28}$$

The observability matrix is

$$O_o = \begin{bmatrix} C_0 \\ C_0 A_0 \\ \vdots \\ C_0 A_0^{\,n-1} \end{bmatrix} \tag{6.29}$$

and the transformation matrix is

$$T = O^{-1}O_0 \tag{6.30}$$

Once we know T, we can find B_o:

$$B_o = C_c^T = T^{-1}B \qquad (6.31)$$

Example 6.9

For the discrete-time system

$$x(k+1) = \begin{bmatrix} 0 & 3 \\ 0 & 4 \end{bmatrix} x(k) + \begin{bmatrix} 1 \\ 0 \end{bmatrix} r(k)$$

$$c(k) = \begin{bmatrix} 1 & 1 \end{bmatrix} x(k)$$

the characteristic equation is

$$\det[zI - A] = z^2 - 4z$$

Therefore

$$A_o = \begin{bmatrix} 0 & 0 \\ 1 & 4 \end{bmatrix} \text{ and } C_o = \begin{bmatrix} 0 & 1 \end{bmatrix}$$

and

$$O_o = \begin{bmatrix} C_o \\ C_o A_o \end{bmatrix} = \begin{bmatrix} 0 & 1 \\ 1 & 4 \end{bmatrix}$$

From Equation 6.30, the transformation matrix is

$$T = O^{-1}O_0 = \begin{bmatrix} C \\ CA \end{bmatrix}^{-1} \begin{bmatrix} 0 & 1 \\ 1 & 4 \end{bmatrix} = \frac{1}{7}\begin{bmatrix} 7 & -1 \\ 0 & 1 \end{bmatrix}\begin{bmatrix} 0 & 1 \\ 1 & 4 \end{bmatrix} = \frac{1}{7}\begin{bmatrix} -1 & 3 \\ 1 & 4 \end{bmatrix}$$

and

$$T^{-1} = \begin{bmatrix} -4 & 3 \\ 1 & 1 \end{bmatrix}$$

From Equation 6.31

$$B_o = T^{-1}B = \begin{bmatrix} -4 \\ 1 \end{bmatrix}$$

and the resulting observability canonical form (OF) is

$$x_o(k+1) = \begin{bmatrix} 0 & 0 \\ 1 & 4 \end{bmatrix} x_o(k) + \begin{bmatrix} -4 \\ 1 \end{bmatrix} r(k)$$

$$c(k) = \begin{bmatrix} 0 & 1 \end{bmatrix} x_o(k)$$

Since the controllability canonical form and the observability canonical form are so closely related as indicated by Equations 6.27, 6.28, and 6.31, we can construct the resulting controllability canonical form as follows:

$$x_c(k+1) = \begin{bmatrix} 0 & 1 \\ 0 & 4 \end{bmatrix} x_c(k) + \begin{bmatrix} 0 \\ 1 \end{bmatrix} r(k)$$

$$c(k) = \begin{bmatrix} -4 & 1 \end{bmatrix} x_c(k)$$

Example 6.10

A discrete system has the following dynamical equations:

$$e(k+1) = \hat{x}(k+1) - x(k+1) = N\hat{x}(k) - Ax(k) + Mc(k) + (L-B)r(k)$$

$$c(k) = \begin{bmatrix} 1 & 1 \end{bmatrix} x(k)$$

Checking controllability and observability, we get

$$V = \begin{bmatrix} B & AB \end{bmatrix} = \begin{bmatrix} 1 & 1 \\ 0 & 0 \end{bmatrix} \text{ and } O = \begin{bmatrix} C \\ CA \end{bmatrix} = \begin{bmatrix} 1 & 1 \\ 1 & 7 \end{bmatrix}$$

and we note that the system is observable but not controllable. If we cast the system into the CF, we would note that that representation was controllable but not observable. If we cast the system into the OF, we would note that that system was, like the original system, observable but not controllable. Now compute the transfer function

$$T(z) = C(zI - A)^{-1}B = \begin{bmatrix} 1 & 1 \end{bmatrix} \begin{bmatrix} z - 1 & -3 \\ 0 & z - 4 \end{bmatrix}^{-1} \begin{bmatrix} 1 \\ 0 \end{bmatrix} = \frac{z - 4}{(z - 4)(z - 1)}$$

and we note the pole-zero cancellation that can be effected. This in fact illustrates a rather general result: If a system is either controllable and not observable or observable and not controllable, then a pole-zero cancellation is possible. Even if it is possible, however, we may not want to cancel a pole and zero out of the transfer function. Caution is advised. The pole and zero may be close but not exact and if they are, for instance, outside the unit circle, we may still have an unstable system even though the simplified transfer function does not indicate it.

Since most well-formed systems are both controllable and observable, let us assume for the remainder of this discussion that they are both. Returning to the idea of the state feedback or pole-placement problem, let us look at the issue in a more realistic fashion. Because it is difficult or expensive to obtain all the system's states by direct measurement, estimators are often designed to estimate the state vector x.

Estimators

Estimators, sometimes known as *observers*, are quite commonly employed in systems that have been designed using the pole- placement method. For pole placement we generally need all the states, all the elements in the state vector $x(k)$. Without direct information about the system states, we can design a system that uses the original system's inputs and outputs as its inputs and produces as output an estimate of the original system's states. A block diagram of an estimator is presented in Fig. 6.8.

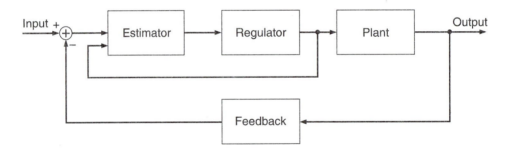

Figure 6.8   Estimator used in discrete-time control system.

The estimator is a dynamic discrete system with state variables $\hat{x}(k)$ and inputs $c(k)$ and $r(k)$. We use $\hat{x}$ as our notation for the estimate of the state vector. Design of an estimator requires the following steps:

1. Write the discrete state equation for the estimator as follows

$$\hat{x}(k+1) = N\hat{x}(k) + Mc(k) + Lr(k) \tag{6.32}$$

2. Since we want $\hat{x}(k) \approx x(k)$, define the error as

$$e(k) = \hat{x}(k) - x(k) \tag{6.33}$$

and we can write

$$e(k+1) = N\hat{x}(k) - Nx(k) = Ne(k) \tag{6.34}$$

3.   Let $c(k) = Cx(k)$ from the original output equation and collect terms; then if we set

$$MC - A = -N$$
$$L - B = 0 \tag{6.35}$$

we end up with the dynamic autonomous error equation

$$e(k+1) = N\hat{x}(k) - Nx(k) = Ne(k) \tag{6.36}$$

whose characteristic equation is given by

$$\det(zI - N) = \det(zI - A + MC) = 0 \tag{6.37}$$

4. Choose the desired pole locations $z = q_1, q_2, \ldots, q_n$ for the error equation so that any initial conditions on the error function will quickly decay to zero. Let the desired error characteristic equation be

$$\alpha_o(z) = (z - q_1)(z - q_2)\ldots(z - q_n) \tag{6.38}$$

and the problem reduces to equating Equations 6.37 and 6.38 and solving for the elements of the M matrix.

Note that this error characteristic equation is the same as the observer characteristic equation. Note also that this equation is very similar to the characteristic equation we presented for the pole-placement problem. Here we need to determine the matrix M and there we needed the state feedback gain matrix K. The state-variable feedback problem and the observer problem are sometimes said to be duals of each other. They have several other similarities that are beyond our present scope. The main issue here is how to determine M. The basic idea is that we want the eigenvalues of N to be close to the origin in the z-plane so that initial conditions on the error will be driven to zero as fast as possible.

Example 6.11

Consider the following system in the observability canonical form and design an observer for it:

$$A = \begin{bmatrix} 0 & 0 & -0.5 \\ 1 & 0 & -0.5 \\ 0 & 1 & -1 \end{bmatrix}$$

$$B = \begin{bmatrix} 1 \\ 2 \\ -1 \end{bmatrix}$$

$$C = \begin{bmatrix} 0 & 0 & 1 \end{bmatrix}$$

If we set $|zI - A + MC| = |zI - N|$ and let

$$M = \begin{bmatrix} m_1 \\ m_2 \\ m_3 \end{bmatrix}$$

and select the three eigenvalues of N to be all at -0.1, we can write

$$\begin{vmatrix} z & 0 & 0.5+m_1 \\ -1 & z & 0.5+m_2 \\ 0 & -1 & z+1+m_3 \end{vmatrix} = |zI - N| = (z+0.1)^3$$

or

$$z^3 + z^2(m_3+1) + z(m_2+0.5) + (m_1+0.5) = z^3 + 0.3z^2 + 0.03z + 0.001$$

and equating coefficients, we get

$$M = \begin{bmatrix} m_1 \\ m_2 \\ m_3 \end{bmatrix} = \begin{bmatrix} -0.499 \\ -0.470 \\ -0.700 \end{bmatrix}$$

Just as the pole-placement problem is very easy to solve when the system we start with is in the controllability canonical form, so is the observer problem quite simple when the system we start with is in the observability canonical form. But generally the design of an estimator/observer involves the solution of n equations in n unknowns. Again, however, there is an Ackermann formula that will simplify our task.

Another Ackermann Formula

Since computing the M matrix can be a tedious task, especially for higher-order systems, we often employ a procedure due to Ackermann. Consider the desired estimator characteristic Equation 6.38, repeated here:

$$\alpha_o(z) = (z - q_1)(z - q_2)...(z - q_n) .$$

Since the Cayley-Hamilton theorem says that every A matrix satisfies its characteristic equation, this equation should be zero if we substitute the closed-loop observer A matrix in place of z. Let that closed-loop A matrix be N where by rearranging Equation 6.35, we can write

$$N = A - MC \qquad\qquad (6.39)$$

In other words, from the Cayley-Hamilton theorem

$$\alpha_o(N) = 0 \qquad (6.40)$$

But for the Ackermann formula we actually need

$$\alpha_o(A) = A^n + \alpha_{n-1}A^{n-1} + \cdots + \alpha_1 A + \alpha_0 I \qquad (6.41)$$

which is generally not equal to zero and where A is the open-loop system matrix of the state equation. Then the formula derived by Ackermann for the estimator gain matrix M is

$$M = \alpha_o(A)\begin{bmatrix} C \\ CA \\ \vdots \\ CA^{n-1} \end{bmatrix}^{-1}\begin{bmatrix} 0 \\ 0 \\ \vdots \\ 1 \end{bmatrix} \qquad (6.42)$$

where n is the order of the system. The middle term is recognizable as the observability matrix O, so we can write

$$M = \alpha_o(A)O^{-1}\begin{bmatrix} 0 \\ 0 \\ \vdots \\ 1 \end{bmatrix} \qquad (6.43)$$

Example 6.12

Solve Example 6.11 using the Ackermann formula:

Solution:

$$A = \begin{bmatrix} 0 & 0 & -0.5 \\ 1 & 0 & -0.5 \\ 0 & 1 & -1 \end{bmatrix}$$

$$B = \begin{bmatrix} 1 \\ 2 \\ -1 \end{bmatrix}$$

$$C = \begin{bmatrix} 0 & 0 & 1 \end{bmatrix}$$

and we need to compute an inverse and do matrix multiplications. We use MATRIX$_X$ for these tasks and get

$$M = (A + 0.1I)^3 \begin{bmatrix} C \\ CA \\ CA^2 \end{bmatrix}^{-1} \begin{bmatrix} 0 \\ 0 \\ 1 \end{bmatrix} = \begin{bmatrix} -0.499 \\ -0.470 \\ -0.700 \end{bmatrix}$$

the same values we had from Example 6.11.

Now the estimator generates estimates of the state variables. Once these are available, we can feed them back through the gains in the K matrix and proceed with the design of the pole-placement problem. The next example will tie these ideas together.

Example 6.13

Let us design a state-variable feedback gain matrix for Example 6.12 that places the poles at $z = 0$, -0.5, 0.5 or $\alpha_d(\lambda) = z(z - 0.5)(z + 0.5)$. Then construct a block diagram that shows how the observer and the state feedback construction relate.

Solution: The original A matrix has eigenvalues at $z = -1.0$ and $\pm 0.707j$ which indicates a marginally stable system. We check controllability because the given system is in the OF, not the CF.

$$V = \begin{bmatrix} B & AB & A^2B \end{bmatrix} = \begin{bmatrix} 1.0 & 0.5 & -1.5 \\ 2.0 & 1.5 & -1.0 \\ -1.0 & 3.0 & -1.5 \end{bmatrix}$$

whose determinant is -8.5, indicating a controllable system. Then the first Ackermann formula yields

$$K = \begin{bmatrix} 0 & 0 & \cdots & 1 \end{bmatrix} \begin{bmatrix} B & AB & \cdots & A^{n-1}B \end{bmatrix}^{-1} \alpha_d(A)$$

$$K = \begin{bmatrix} 0 & 0 & \cdots & 1 \end{bmatrix} V^{-1} \alpha_d(A)$$

and we need the characteristic equation

$$\alpha_d(A) = A(A - 0.5I)(A + 0.5I)$$

Using MATRIX$_X$ again for the inversion and matrix multiplication, we get

$$K = \begin{bmatrix} 0 & 0 & 1 \end{bmatrix} V^{-1} \{A(A - 0.5I)(A + 0.5I)\} = [0.1912 \ -0.4559 \ 0.2794]$$

A block diagram relating estimator and state-variable feedback is shown in Fig. 6.9.

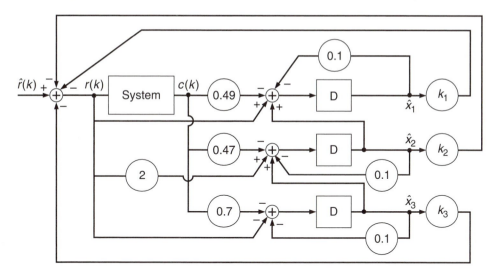

Figure 6.9   Block diagram for Example 6.13.

In a final example in this section, drawn from the work of van der Kruk and Scannell, as reported in Reference [17], we consider a more realistic system: a velocity observer design for a digital motion controller. The control strategies developed are implemented on the MCV60 motion control board whose hardware is based around the TMS320C25 digital signal processor running at 40 Mhz.

Example 6.14[1]

Development of a digital motion controller must take account of the sampled data nature of the system. For example, a stable-position servo system must provide electronic damping, which often means tachometer feedback or else a simple derivative action in the position feedback loop to generate velocity. Position and velocity are required state variables, but only position is directly available. Using a tachometer increases the cost of the servo system, often prohibitively, whereas a simple digital differentiation technique amplifies the quantization noise on the digital position signal. This causes excessive current ripple in the motor, together with unpleasant audible noise. Faster sampling can often minimize the noise problem, but again, at prohibitive cost. The design presented in this example uses a velocity observer to drastically reduce the quantization problem associated with simple digital velocity estimators.

   Introduction of the microprocessor, and more recently the signal processor, has radically altered the field of high-performance servo control over the past decade. The advent of digital techniques has presented the designer with tremendous flexibility in control algorithm design. In addition, the provision of extensive diagnostics and status information has become a relatively simple operation, thus easing the tasks of system development and support. However, the migration from analog to digital has several associated problems. In particular, the design of the control algorithm must take account of the sampled data nature of the system. Problems due to the delays introduced by the sample period and the calculation time must be considered carefully in the design of the feedback gain parameters. The quantization noise due to the digital nature of the position information must also be analyzed carefully and its effects minimized.

   Consider the simple block diagram of the digital servo system in Fig. 6.10.

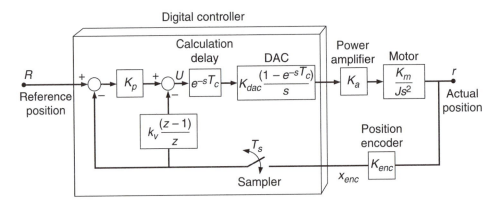

Figure 6.10   Simple second-order digital controller.

[1] Portions of this example reprinted with permission from PCIM. Portions reprinted with permission from ZM Communications GmbH. See Reference [17].

Assume that the power amplifier has a large bandwidth compared with the servo loop and may therefore be modeled as a gain element. The motor model is a double integrator and neglects friction and mechanical resonances.

The open-loop transfer function of the continuous elements, including the sample-and-hold effect and the calculation delay is

$$H(s) = \frac{X_{enc}(s)}{U(s)} = K \frac{e^{-sT_C}(1 - e^{-sT})}{s^3}$$

where

$$K = K_{dac}K_a K_m K_{enc} / J$$

K_{dac} is the gain of the D/A converter, K_a is the gain of the amplifier, K_m is the motor constant, K_{enc} is the resolution of the position converter, and J is the motor inertia.

The most convenient method of analyzing this sampled-data system is to convert H(s) into its discrete-time equivalent, H(z). However, the z-plane analysis only provides information at the sample instants (i.e., fractional delays are not allowed). Thus, to examine the effect of the calculation delay, T_C, the two extreme cases are considered: no calculation delay, $T_C = 0$, and maximum calculation delay, $T_C = T_S$.

Calculation of the feedback parameters is first considered for zero calculation delay. For $T_C = 0$ we have

$$H(z)|_{T_C=0} = Z\{L^{-1}[H(s)]\} = \frac{1}{2} KT_s^2 \frac{(z+1)}{(z-1)^2}$$

A suitable value of the velocity feedback gain, K_V, is calculated by considering the loop transfer function. From Fig. 6.10, the motor velocity is approximated by using the pulse count technique, which is the backward rectangular rule for representing the differentiation operation. The open loop transfer function of the velocity loop is

$$V(z)|_{T_C=0} = \frac{z-1}{z} H(z)|_{T_C=0} = \frac{1}{2} KT_s^2 \frac{(z+1)}{z(z-1)}$$

Using the root-locus technique, a suitable value of K_V may be derived. A robust selection, giving sufficient design freedom for the outer position loop, is $K_v = 0.343/KT_s^2$. This gives a damping ratio of 1.0. Using the value of K_V, the open-loop transfer function of the position loop is

$$P(z)|_{T_C=0} = \frac{1}{2}KT_S^{\,2}\frac{z(z+1)}{(z-1)(z-0.414)^2}$$

As before, the root-locus technique is used to calculate a value for the position feedback gain, K_P. Selecting a damping ratio of 0.7 gives $K_P = 0.072 / KT_S^2$.

In a similar manner, the loop parameters may be determined when a calculation delay of one sample period is assumed. In this case we can write

$$H(z)|_{T_C=T_S} = \frac{1}{2}KT_S^2\frac{(z+1)}{z(z-1)^2}$$

The corresponding values of K_V and K_P for the same damping ratios used in the zero-delay case are $K_V = 0.180 / KT_S^2$ and $K_P = 0.018 / KT_S^2$.

The results obtained are summarized in Table 6.1, where F_S is the sample frequency $(F_s = 1/T_s)$.

Table 6.1   Velocity and Position Feedback Gains

	$K_V.KT_S^2$	$K_p.KT_S^2$	Total Delay	Bandwidth
$T_C = 0$	0.343	0.072	T_S	$F_S/17.5$
$T_C = T_S$	0.180	0.018	$2T_S$	$F_S/35.0$

These results have also been verified for fractional calculation delays (i.e., $0 < T_C < T_S$). The general conclusion is that for a good design the sample frequency must be at least 17.5 times higher than the required bandwidth of the position loop.

The straightforward pulse count method of velocity estimation employing the backward rectangular approximation for differentiation results in poor error resolution at low speeds. The quantization error in the velocity signal becomes worse with increasing sample frequency and can cause excessive current ripple in the motor together with audible noise. This problem can be reduced by using a position encoder with a greater resolution, but this is a rather expensive solution. In addition, increasing encoder resolution unnecessarily increases the data speed, which can lead to a decrease in the maximum servo velocity. A velocity observer estimating the servo velocity with a higher resolution than the pulse count method can be used to overcome this quantization problem. The discrete-time transfer function from the servo command, U (z), to the velocity output, $X_{enc}(z)$, is given by

$$H(z)|_{T_C = T_S} = \frac{1}{2} KT_S^2 \frac{(z+1)}{z(z-1)^2} \quad \text{with } T_C = T_S$$

The transfer function from the servo command, U(z), to the velocity output, is

$$G(z) = \frac{V(z)}{U(z)} = Z\{L^{-1}[sH(s)]\} = KT_S^2 \frac{1}{z(z-1)}, \quad T_C = T_S$$

These equations yield the observer structure shown in Fig. 6.11, where

$$K_{dc} = KT_S^2$$

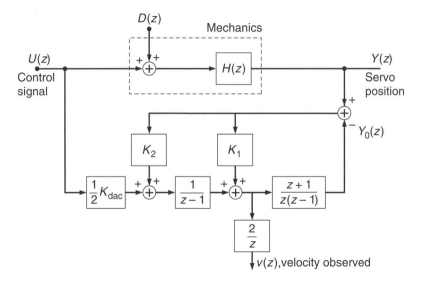

Figure 6.11   Velocity observer block diagram.

Two feedback terms are used to correct for deviations between the observer and the system. The choice of K_1 and K_2 involves a trade-off between the bandwidth of the observer correction loop and the quantization of the estimated velocity, V(z). Since the objective of the velocity observer is to reduce the velocity quantization level, the choice of K_1 and K_2 are determined using this criterion. The value of K_1 must be less than 1/2 to provide a better resolution than the pulse count method. Practical values are $K_1 = 0.04$ and $K_2 = K_1^2$ resulting in a resolution enhancement factor of $\frac{1}{2}K_1 = 12.5$ and an observer 3db bandwidth of $\frac{F_s}{105}$.

The performance improvement yielded by the velocity observer is demonstrated in Fig. 6.12. A reference velocity signal of 2.5 position increments per sample period $2.5\mathrm{inc}/T_s$ is applied to a brushless linear motor used in the Philips chip mounting machines. Fig. 6.12a shows the current in the motor when the pulse count method of velocity estimation is used. In contrast, when the velocity is estimated using the observer, the high-frequency current components are eliminated as shown in Fig. 6.12b. This reduction in the current ripple results in less motor heating, less dissipation in the power amplifier, and a significant reduction in the audible noise level.

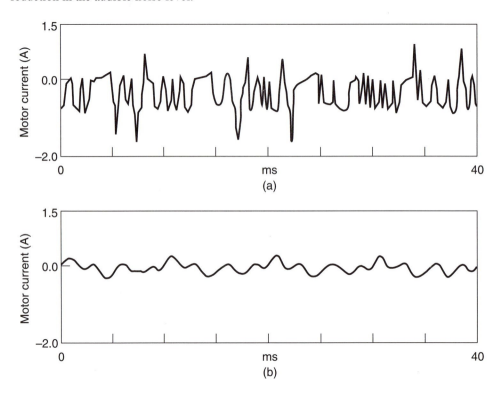

Figure 6.12 Performance improvement yielded by a velocity observer: (a) pulse count; (b) velocity observer.

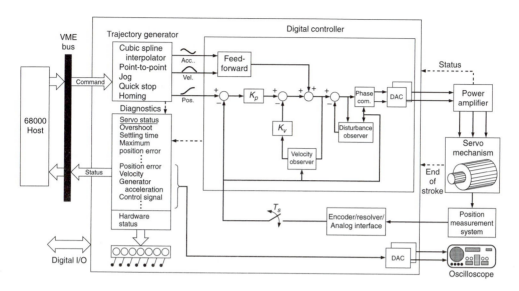

Figure 6.13   Block diagram of the MCV60 connector to a three-phase brushless motor.

The MCV60 hardware is based on the TMS320C25 signal processor running at 40 MHz. This high clock frequency, together with the arithmetic capabilities of the signal processor, yields a high sample frequency and a short calculation delay (Ts = 150 μsec, Tc = 40 μsec). The maximum encoder data speed is 7 Mhz. On-board memory includes 16K words of program memory and 8K words of data memory, of which 2K words is in dual-ported RAM. This provides high speed, bi-directional communication with the VME host computer. The data RAM has a battery back-up for the retention of system parameters when the card is not powered up. Three different position encoders are supported. An incremental encoder interface is standard on the card, while piggyback interfaces for resolvers and sine wave encoders may be simply mounted. Two 16-bit DACs deliver drive signals to the current amplifiers, while a second pair of DACs provide monitor information. Hardware synchronization between several cards is also provided with the facility for master-slave control. In addition, five optically isolated outputs and four optically isolated inputs are provided for interfacing with programmable logic controllers.

The motion control software may be divided into card and host levels. At the card level, the controller algorithm implements the various strategies already described. In addition, continuous path movements can be implemented using cubic spline interpolation techniques. Software for auto-homing is provided and, for brushless motors, an automatic magnetic alignment routine together with commutation software is available. Extensive hardware and servo diagnostics are performed at power-up, and critical hardware checks are performed each sample period. System monitoring is performed at each sample period generating the two monitor DAC variables. Other performance indicators are also recorded, such as the maximum error during a movement, the overshoot, the setting time, and so on.

At the host level, drivers are provided for communicating with the MCV60. A menu-driven, user-friendly test environment initializes and tunes system parameters. Self-tuning facilities initialize the controller parameters to suitable values based on the system characteristics. Most refined parameter tuning may then be simply carried out by a series of well-defined tests. Communication with the card occurs each time that a parameter is changed. This facilitates on-the-fly variation of system parameters for use in systems employing gain-scheduling techniques.

Fig. 6.14 and 6.15 demonstrate the speed and accuracy of the MCV60 motion controller. In Fig. 6.14, the response of a brushless linear motor to a set point displacement of 10mm (10,000 increments) is shown. Fig. 6.15 shows the motor velocity profile and the position error of the same servo when running at maximum velocity (data speed = 1.5 Mhz). The maximum position error is only 12 increments (12 μm).

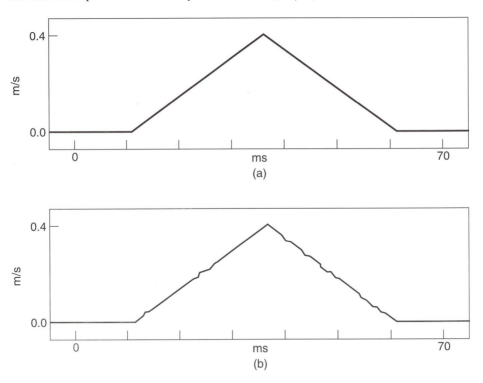

Figure 6.14   Profiling accuracy for a displacement of 10 mm in 50 ms: (a) generator velocity; (b) observer velocity.

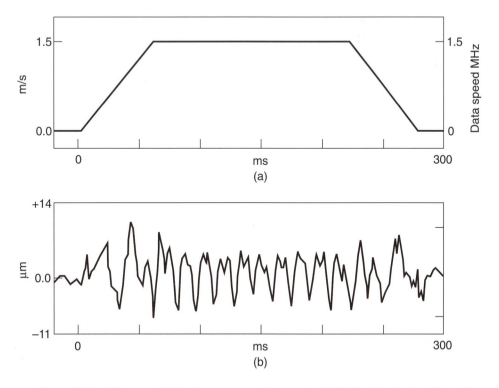

Figure 6.15   High-speed following accuracy; data speed = 1.5 MHz: (a) velocity; (b) position
error.

6.4 LINEAR QUADRATIC OPTIMAL DESIGN

In the pole-placement problem the location of the closed-loop poles in the z-plane was at
issue. If the system we are working on is essentially second order, the pole locations can be
selected to achieve certain desired damping ratios and/or undamped natural frequencies and
these, in turn, translate into various transient specifications. A different approach to pole-
placement design is to select closed-loop poles such that a cost function is minimized.
Several different kinds of cost functions have been considered, but the most popular cost
function is the quadratic cost, an infinite series consisting of the weighted sum of the
squares of the control variable and the states. The weighting terms on the control and on the
states are what the designer adjusts, in a trial-and-error fashion, until specifications are met.

   Let us assume that we want to minimize with respect to r(k) the following cost
functional (a functional is a function of a function):

$$J = \sum_{k=0}^{\infty}\{x^T(k)Qx(k) + r^T(k)Rr(k)\} \tag{6.44}$$

subject to the constraint of the state equation

$$x(k+1) = Ax(k) + Br(k) \tag{6.45}$$

The matrices Q and R are assumed to be positive definite. In the case of SISO systems, which is our primary concern, Q is $n \times n$ and R is a scalar.

Since we are dealing with a functional, the methods of simple calculus will not do. The calculus of variations method, the Lagrange multiplier approach, Bellman's principal of optimality, and the Pontryagin minimum Principle are various techniques that have been employed to solve this problem. Applying these methods results in the need to solve the famous Riccati equation, which is generally a nonlinear matrix difference equation that can only be solved numerically. Since the summation in Equation 6.44 is infinite, our problem is actually a little simpler. The Riccati equation is a non-linear difference equation if the summation is finite, but reduces to a matrix algebraic equation in our case of an infinite sum.

The solution to the minimization problem is exactly of the same form as the pole-placement problem:

$$r(k) = -Kx(k) \tag{6.46}$$

where the gain K is itself time-varying if the summation in Equation 6.44 is finite, but is a constant in the case at hand with an infinite sum. We assume here that the reference input added to the right-hand side of Equation 6.46 is zeroed out; this is the *regulator problem*. The idea is to keep the output close to zero for all initial conditions that might appear in the system.

Our approach to this LQR (linear quadratic regulator) problem will be to sidestep the mathematics and rely on CAE tools. In fact, both $MATRIX_X$ and MATLAB have solutions to the LQR problem. We look at the MATLAB statement DLQR.

Example 6.15

Consider the system of Example 6.13 and design it using Q = I, a 3×3 identity matrix, and R = 1 in the LQR:

$$A = \begin{bmatrix} 0 & 0 & -0.5 \\ 1 & 0 & -0.5 \\ 0 & 1 & -1 \end{bmatrix}$$

$$B = \begin{bmatrix} 1 \\ 2 \\ -1 \end{bmatrix}$$

$$C = \begin{bmatrix} 0 & 0 & 1 \end{bmatrix}$$

The MATLAB statement is [K, S, E]=DLQR (A, B, Q, R). Note that the solution does not depend on the C matrix. But we will calculate $c(k) = Cx(k) = x_3(k)$ and plot this output response for the case of zero reference input and initial conditions

$$x(0) = \begin{bmatrix} 1 \\ 1 \\ 1 \end{bmatrix}$$

With Q = I and R = 1.0 we get the optimal feedback gain matrix K=[0.2796 -0.2901 0.0974]. Also returned with the MATLAB DLQR statement is the S matrix, which is the solution of the algebraic Riccati equation, and the E vector, which is the vector of the closed-loop eigenvalues. The eigenvalues in this case, for instance, are $-0.0541 \mp j0.2065$ and -0.4939. The plot of $c(k) = Cx(k) = x_3(k)$ appears in Fig. 6.16.

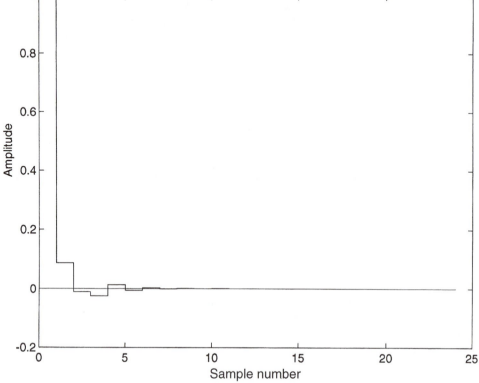

Figure 6.16   Optimal response for the linear quadratic regulator.

For comparison purposes we can plot the response of the system designed via pole placement when we arbitrarily placed the poles at -0.5, +0.5, and 0.0. That response appears in Fig. 6.17. Note that the optimal response is slightly smoother. An even smoother response is attainable if we vary the Q and R weighting terms. For example, changing the Q to 5I instead of just I yielded the output response shown in Fig. 6.18.

In this case we have K=[0.2875   -0.3008 0 0.996] and closed-loop eigenvalues $-0.0773 \pm 0.0706\text{j}$ and -0.4317. As opposed to the pole-placement problem, the optimal linear regulator problem allows the designer to have a rationale for setting the closed-loop poles; namely, these are the ones that minimize such and such a cost function. We turn now to a brief consideration of fuzzy logic control.

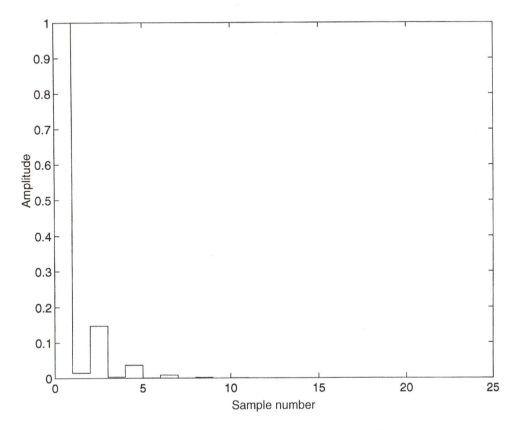

Figure 6.17   Output response with poles placed at -0.5, 0.5, and 0.0.

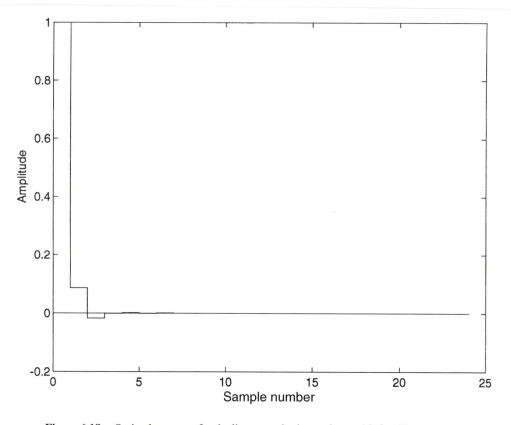

Figure 6.18   Optimal response for the linear quadratic regulator with Q = 5I.

6.5 FUZZY LOGIC CONTROL

Fuzzy logic control is a technique that is growing in popularity among control systems engineers. There are many areas of uncertainty or fuzziness in real physical systems, and an efficient way of dealing with this fuzziness is by the mechanism of fuzzy logic. Fuzzy logic controllers (FLCs) are increasingly being applied to systems with unsharp boundaries that do not lend themselves to explicit difference or differential equation descriptions. FLCs provide adequate control which is inexpensive, easy to implement, and robust. In this section we discuss some of the key ideas associated with FLCs and present two examples to illustrate these ideas.

Fuzzy logic control is generated in a three-step process: fuzzification, inference, and defuzzification. Think of an FLC as a block like a PID controller block which has an input and an output and in between there is a system that performs some kind of control activity. The input to the FLC is typically a signal or signals, position and velocity, for example, which are variables in a real physical system available as measured deterministic or "crisp" signals. In *fuzzification* crisp inputs are mapped into membership functions of a fuzzy set. The fuzzy set consists of the input values and their corresponding membership function: $F = \{x, f(x) / x \varepsilon X\}$. The membership function $f(x)$ maps each element of X to a membership value between 0 and 1. For example, assume that the input is a voltage v, which ranges from -15 to +15 volts. If –15 to 0 volts is considered "low," and 0 to +15 volts is considered "high," then perhaps -7.5 to +7.5 volts can be considered "medium" and the voltage might be mapped into a membership function as indicated in Fig. 6.19.

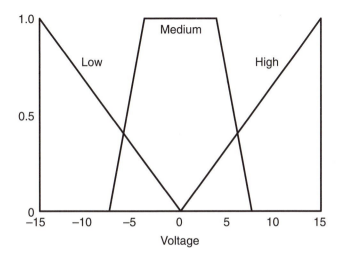

Figure 6.19   Relative membership in a fuzzy set.

A voltage of 3 volts, for instance, has a relative membership of 1.0 in the function MEDIUM and 0.2 in the function HIGH. A key idea here is that a fuzzy logic system can respond based not only on membership in a fuzzy set but also on the *degree* of such membership.

In the *inference* stage of the FLC, we develop control rules to relate fuzzified inputs to fuzzified outputs. (Then these outputs are defuzzified and turned into crisp variables that can drive real physical systems.) Several rules can be generated to define the control. For example:

1.   If voltage is low, set motor speed to slow.
2.   If voltage is medium, set motor speed to medium.
3.   If voltage is high, set motor speed to fast.

Typically, these inference rules will be of the form of if-then statements. We may have multiple inputs and have rules like "if a and b and c, then d." Because inference rules are based on ordinary language expressions, not on mathematical definitions, any linguistically expressible relationship – even if the systems being controlled are nonlinear or without solid mathematical representations of any kind – can be defined in an FLC. Also, we can give different rules relative weightings so that they have different influences on the control action.

After the FLC fuzzifies the inputs and applies the inference system to them, it must generate an output that is in appropriate form. For example, if the voltage was of a nature that implied a motor speed of SLOW with a relative weighting of 0.75 and MEDIUM with a relative weighting of 0.50, we need to convert that fuzzy speed into a crisp number, such as 60 rpm, which can be employed in the physical system that is being controlled. The process of converting a fuzzy output of an inference system to a crisp form is called *defuzzification*. Two common methods of defuzzification are the *maximum defuzzification method* and the *centroid calculation defuzzification* method. We consider only the latter. What the centroid calculation produces is the center of the area under a curve. That curve is obtained from aggregating all the outputs of the implication process to yield a single fuzzy set for the overall FLC output.

MATLAB has a Fuzzy Logic Toolbox [14,15] that greatly eases the burden of generating FLCs. In fact, combined with SIMULINK, the FUZZY LOGIC Toolbox enables the designer to generate an FLC and embed it directly into a simulation and to adjust parameters empirically to design a fuzzy logic controlled system. We illustrate the use of these tools in the following example, which for the sake of simplicity is continuous time rather than discrete time in nature. FLCs work equally well with both kinds of systems.

Example 6.16

Consider the system with unity gain feedback presented in Fig. 6.20. With no feedback the system oscillates in response to a unit step because the open loop transfer function has poles on the $j\omega$ -axis of the s-plane. With unity gain feedback the highly underdamped step response appears as in Fig. 6.21. For comparison purposes we employ a PID feedback controller in place of the unity gain feedback as illustrated in Fig. 6.22.

With the derivative term zero (the PI controller) the step response is unstable. But the PD controller with proportional and derivative gains of one and the PID controller with all gains equal to 1 yield nice stable responses, as illustrated in Fig. 6.23 and 6.24. Now we replace the PID block with an FLC as indicated by the SIMULINK block diagram in Fig. 6.25.

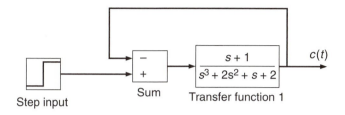

Figure 6.20   SIMULINK block diagram of system for Example 6.16.

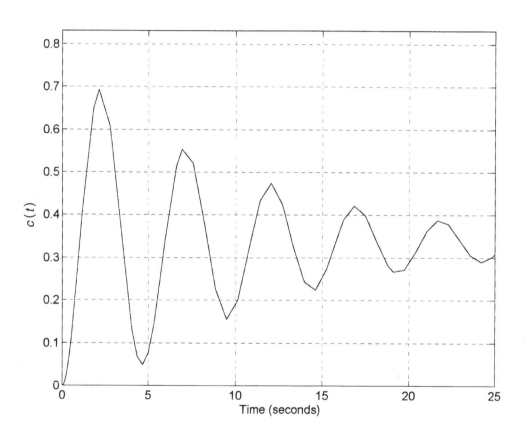

Figure 6.21   Step response with unity gain feedback.

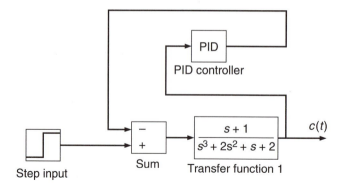

Figure 6.22   PID feedback controller.

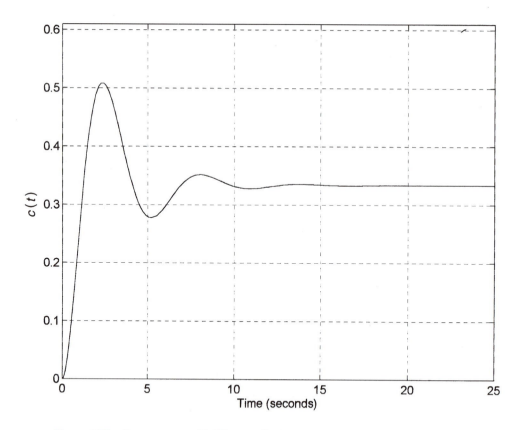

Figure 6.23   Step response with PD controller.

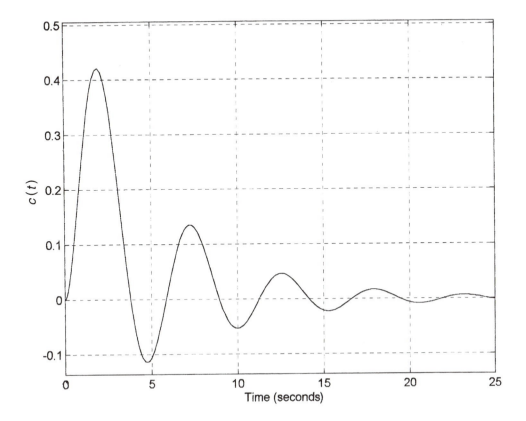

Figure 6.24   Step response with PID controller.

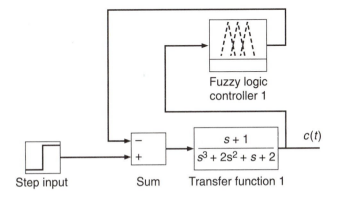

Figure 6.25   FLC in the feedback path.

In the Fuzzy Logic Toolbox the FLC can be designed with the Fuzzy Inference System Editor. The FIS Editor allows the designer to fuzzify the input variables by defining membership functions, generate the fuzzy inference rules, and define membership functions for the output that is to be defuzzified. We tried various types of membership functions but settled on the trapezoidal functions for both input and output (Fig. 6.26). The inference rules we chose were:

1. If input is low, output is low.
2. If input is medium, output is medium.
3. If input is high, output is high.

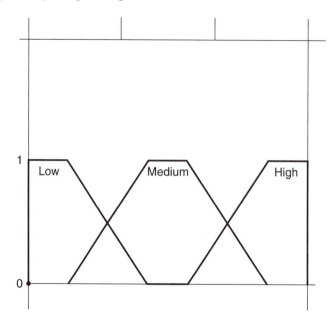

Figure 6.26   Typical trapezoidal membership functions.

The results were very dependent on the ranges we chose for the input and the output. With the assumption that the input varied from -2 to +2 and the output did the same, we got a step response as in Fig. 6.27, which is even more oscillatory than the unity gain control case.

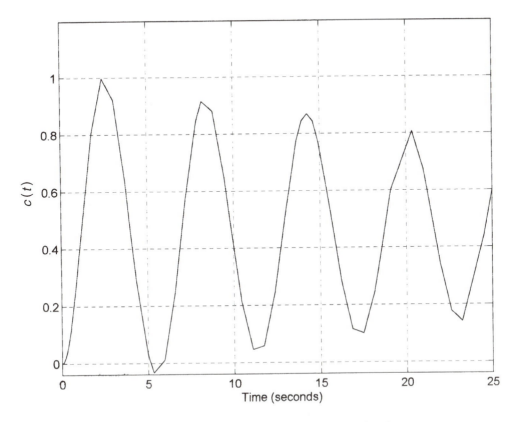

Figure 6.27   FLC controller with -2 to +2 input and output variational range.

Other input and output ranges yielded more or less damping and overshoot. With an input range of 0 to 0.7 and an output range of 0 to 2, we got some nice transients (Fig. 6.28). Keeping the input range the same but increasing the output range from 0 to 2.5, 0 to 3.0, 0 to 3.5, and so on, yielded a more damped response and an increased response time. See Fig. 6.29 with output range 0 to 5.5. Going beyond 5.5, however, yielded increased oscillation and eventually instability. Several parameters can be varied like this in a trial-and-error fashion to construct an FLC. Note that the transients provided by the FLC in this problem compared favorably to the PID transients. But, as the next example illustrates, PID control can be more time consuming to develop than an FLC and results in a controller that is less robust than the fuzzy logic control scheme. Also, in this example the system is linear and the plant transfer function is well defined. Fuzzy logic control works best when the systems are complex, nonlinear, or do not have mathematical models or descriptions readily available.

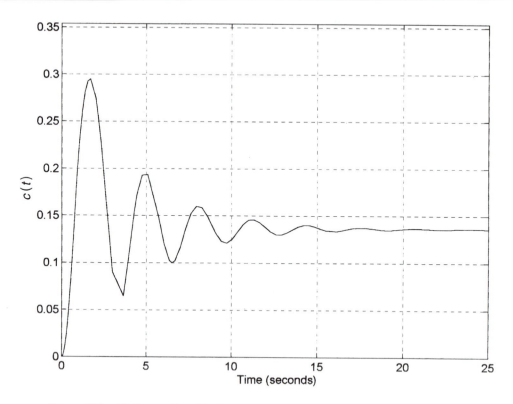

Figure 6.28   FLC controller with 0 to 0.7 input range and 0 to 2 output range.

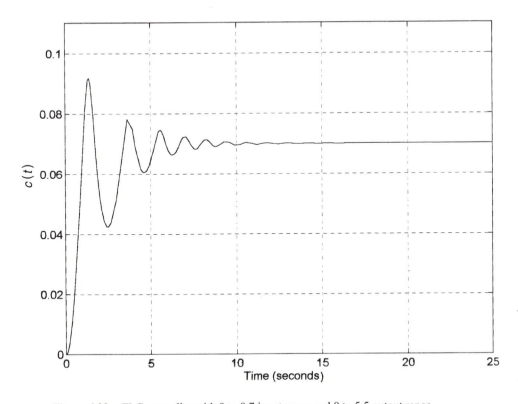

Figure 6.29   FLC controller with 0 to 0.7 input range and 0 to 5.5 output range.

Example 6.17   Fuzzy ServoMotor Control[2]

This implementation, based on Reference 16, employs the Power-14 Board, a programmable PID motor control board that uses a TI TMS320C14 DSP chip. A schematic of the board and motor is presented in Fig. 6.30.

---

[2] Courtesy of Texas Instruments. See Reference [16].

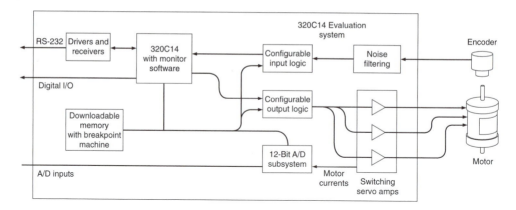

Figure 6.30   Power 14 system.

The PID compensator code uses position and velocity of a motor for inputs and motor input current as the output. To modify the code for fuzzy logic, the PID compensator section of the code is replaced with a fuzzy logic compensator. Then the PID and FLC controls are compared.

For the PID compensator, the proportional, integral, and differential variables are derived from the motor encoder sensor that detects position. The error between position and the desired position is used as the position input for the proportional section of the PID. A backward difference between the error in the current cycle and the error in the previous cycle yields an approximation for velocity and is used as the input for the derivative section of the PID. The input for the integral section of the PID is formed by adding up error (velocity) terms. The PID parameters are set via trial and error.

For the FLC, the PID implementation is removed and fuzzification of the two input variables, position (Theta) and velocity (dTheta), is initiated. Five fuzzy logic ranges (linguistic variables) were chosen for the membership functions and used for both input variables and for the output variable as well. Fig. 6.31 indicates these ranges as overlapping isosceles triangles.

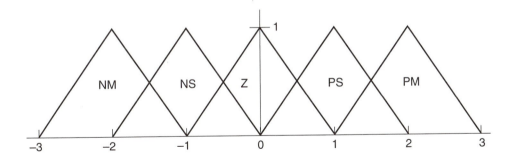

Figure 6.31 Membership functions.

The FLC has one output variable, the motor current. Eleven inference rules, relating the two input variables to the output variable, are listed in Table 6.2. The inference rules are of the form "if a and b, then c." The defuzzification was done by the centroid method. Determining appropriate ranges for the variables involved took most of the effort in this implementation. For more details, see Reference 16.

The fuzzy logic code required about 2000 instruction cycles to execute on the TMS320C14. This implies a 400-μs period because each instruction cycle is 200 ns on the TMS320C14. Since the update period of the motor is 3.8 ms, the FLC is quite adequate. Fig. 6.32 presents the transient response plots for the PID controller.

Table 6.2   List of Inference Rules

If Theta Is:	And dTheta Is:	Then Motor Current Is:
Z	Z	Z
PS	Z	NS
PM	Z	NM
NS	Z	PS
NM	Z	PM
Z	NS	PS
Z	NM	PM
Z	PS	NS
Z	PM	NM
PS	NS	Z
NS	PS	Z

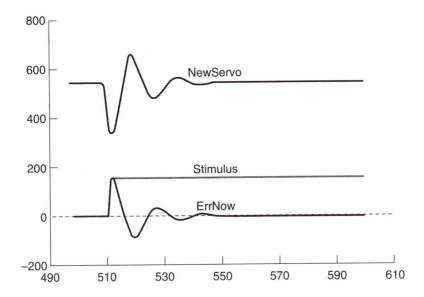

Figure 6.32 Transients for the PID controller.

The scaled output of the PID controller is the variable stored in New Servo. The stimulus is a step input and ErrNow is the difference between actual position and desired position. For the FLC the transients are plotted in Fig. 6.33. In this case the scaled output of the FLC is stored in the NewServo location. Note that the fuzzy transients are actually smoother than the PID transients, but the FLC ErrNow variable does not go to zero. A bias may be acceptable if it is small enough. Or a simple filter circuit of the high-pass variety following the FLC can be used to filter out the bias.

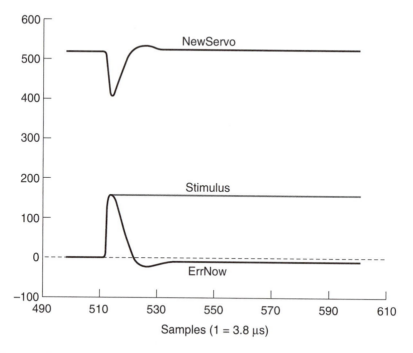

Figure 6.33   Transients for the fuzzy logic controlled system.

6.6 SUMMARY

In this chapter we introduced modern design techniques based on the state-variable formulation of discrete-time linear systems. The pole-placement problem was investigated first and its dependence on the notion of controllability was stressed. The state estimator problem was considered next. That problem's dependence on the idea of observability was discussed. Several examples illustrated not only the pole-placement and estimator problems but their interaction as well.

The basic assumption in pole-placement techniques is that we know the desired location of the poles and we design the system based on that information. But in optimal control theory, the goal is to find the optimum pole locations based on the goal of minimizing a cost function and then design the system using those poles. We illustrated an optimal design using CAE software packages, in particular, the MATLAB software tools. Finally, the chapter concluded with an introductory discussion of fuzzy logic control. Again, MATLAB was used, in particular, their Fuzzy Logic Toolbox in conjunction with SIMULINK, to provide the vehicles for efficient design and testing of a variety of control systems. Fuzzy control turns out to be most effective when the systems to be controlled are complex with perhaps nonlinearities or insufficiently modeled elements.

PROBLEMS

6.1   For the system of Fig. 6.34 determine the constants a and b such that the closed-loop poles are at -0.2 and +0.2.

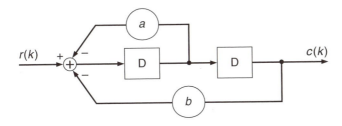

Figure 6.34   System for Problem 6.1.

6.2   For the system of Fig. 6.34, determine the constraints, if any, on a and b such that the system remains controllable.

6.3   Check the controllability of the following state equations:

(a) $x(k+1) = \begin{bmatrix} 1 & 1 \\ 2 & 1 \end{bmatrix} x(k) + \begin{bmatrix} 0 \\ 1 \end{bmatrix} r(k)$

(b) $x(k+1) = \begin{bmatrix} 0 & 0 \\ 0 & 1 \end{bmatrix} x(k) + \begin{bmatrix} 1 \\ 1 \end{bmatrix} r(k)$

6.4   Check the controllability and observability of the following dynamical equations:

$$x(k+1) = \begin{bmatrix} 1 & 1 \\ 2 & 1 \end{bmatrix} x(k) + \begin{bmatrix} 0 \\ 1 \end{bmatrix} r(k)$$

$$c(k) = \begin{bmatrix} 1 & 1 \end{bmatrix} x(k)$$

6.5   Transform the following dynamical equations to the controllability canonical form:

$$x(k+1) = \begin{bmatrix} 0 & 1 \\ 0 & 1 \end{bmatrix} x(k) + \begin{bmatrix} 1 \\ 1 \end{bmatrix} r(k)$$

$$c(k) = \begin{bmatrix} 1 & 1 \end{bmatrix} x(k)$$

6.6   Determine the value of b for which the following system is uncontrollable:

$$x(k+1) = \begin{bmatrix} 0 & 0 \\ 0 & 1 \end{bmatrix} x(k) + \begin{bmatrix} b \\ 1 \end{bmatrix} r(k)$$

$$c(k) = \begin{bmatrix} 1 & 1 \end{bmatrix} x(k)$$

6.7   Use Ackermann's formula to determine an appropriate gain matrix K for the following system if we want poles at -0.1 and +0.1.

$$x(k+1) = \begin{bmatrix} 1 & 0 \\ 1 & 4 \end{bmatrix} x(k) + \begin{bmatrix} 0.5 \\ 0.4 \end{bmatrix} r(k)$$

$$c(k) = \begin{bmatrix} 1 & 1 \end{bmatrix} x(k)$$

6.8 Use Ackermann's formula to determine an appropriate gain matrix K for the following system if we want poles at -0.1, 0.0, and +0.1:

$$x(k+1) = \begin{bmatrix} 1 & 0 & 0 \\ 1 & 2 & 0 \\ 1 & 1 & 1 \end{bmatrix} x(k) + \begin{bmatrix} 0.2 \\ 0.3 \\ -0.2 \end{bmatrix} r(k)$$

$$c(k) = \begin{bmatrix} 1 & -1 & 1 \end{bmatrix} x(k)$$

6.9 Design an observer for the system given in Problem 6.8, checking to make sure that the system is observable, and then construct a block diagram for the observer with the feedback gains calculated in Problem 6.8.

6.10 Determine the optimal control for the system in Problem 6.8. Use the MATLAB or MATRIX$_X$ linear quadratic regulator tools and assume that

$$Q = \begin{bmatrix} 0.5 & 0 & 0 \\ 0 & 0.2 & 0 \\ 0 & 0 & 1.2 \end{bmatrix} \text{ and } R = 1.0.$$

Compare transients with

$$x(0) = \begin{bmatrix} 1 \\ 1 \\ 1 \end{bmatrix}$$

and letting the input be zero.

6.11 Using the Fuzzy Logic Toolbox, construct the FLC for the system in Problem 6.8. Try various combinations to see if you can improve on the results obtained from the optimal control problem. Discuss your results.

6.12 Must a system be controllable if we use the optimal linear quadratic formulation? Check it out. Construct a second-order uncontrollable system and try to control it via the LQR mechanism. Discuss your results.

References

1. C. L. Philips and H. T. Nagle, *Digital Control Systems Analysis and Design*, 2nd ed., Prentice Hall, Upper Saddle River, NJ, 1990.

2. H. F. Van Landingham, *Introduction to Digital Control Systems*, Macmillan New York, 1985.

3. G. F. Franklin, J. D. Powell, and M. L. Workman, *Digital Control of Dynamic Systems*, 2nd ed., Addison-Wesley, Reading, MA, 1990.

4. R. C. Dorf, *Modern Control Systems*, Addison-Wesley, Reading, MA, 1980.

5. C. T. Chen, *Linear System Theory and Design*, CBS College Publishing, New York, 1984.

6. J. A. Cadzow and H. R. Martens, *Discrete-Time and Computer Control Systems*, Prentice Hall, Upper Saddle River, NJ, 1970.

7. C.L. Philips and R.D. Harbor, *Feedback Control Systems*, Prentice Hall, Upper Saddle River, NJ, 1988.

8. *Digital Control Applications with the TMS 320 Family: Selected Application Notes*, Applications Book, Texas Instruments, Dallas, TX, 1991.

9. C. Slivinsky and J. Borninski, *Control System Compensation and Implementation with the TMS32010 Digital Signal Processing Applications with the TMS320 Family: Theory, Algorithms, and Implementations; Application Book Vol. 1*, Texas Instruments, Dallas, TX, 1989.

10. *Digital Control System Design with the ADSP-2100 Family Application Note*, Analog Devices, Inc., Norwood, MA, 1994.

11. *Control Theory, Implementation, and Applications with the TMS320,* Family Seminar Workbook, Texas Instruments, Dallas, TX, 1991.

12. *MATRIXx/System Build Tour Guide*, Integrated Systems, Inc., Santa Clara, CA, 1990.

13. M. O'Flynn, and E. M. Moriarty, *Linear Systems, Time Domain and Transform Analysis*, J. Wiley, New York, 1987.

14. J. S. R. Jang and N. Gulley, *Fuzzy Logic Toolbox,* the Math Works, Inc., Natick, MA, 1995.

15. M. Beale and H. Demuth, *Fuzzy Systems Toolbox*, PWS, Boston, 1994.

16. M. George, Jr., *Implementation of Fuzzy Servo Motor Control on a Programmable TI TMS320C14 DSP*, Application Note, Texas Instruments, Dallas, Tx, 1993.

17. R. Vanderkruk and J. Scannell, *Motion Controller Employs DSP Technology, Intelligent Motion,* PCIM, Sept. 1988.

The MATRIX$_x$ and MATLAB Design and Analysis Software

A.1 MATRIX$_x$[1]

MATRIX$_x$ is a linear systems analysis tool for engineers and scientists. It is an interactive matrix manipulation environment which combines the numerical tools LINPACK and EISPACK with a comprehensive graphics facility and an expandable function library. MATRIX$_x$ is the entry-level tool for the ISI Product Family, which includes comprehensive systems analysis and control design as well as nonlinear simulation, block diagram system modeling, and automatic real-time code generation and implementation.

Control Design Module

The Control Design Module contains a comprehensive set of functions for controls and system engineers. All functions in these MATRIX$_x$ modules can be accessed by typing HELP at the prompt that is displayed after boot-up. HELP will yield a listing of all available functions. HELP followed by, for example, ROOT LOCUS will yield specific details about that specific function. Some of the MATRIX$_x$ features are:

---

1   Section A.1 courtesy of Integrated Systems, Inc. Portions reprinted with permission.

CLASSICAL	Root locus, Bode, Nyquist, Nichols, pole and zero calculations, gain and phase margins
MODERN	Pole placement, LQR, LQG, controllability/observability, solutions to the Riccati equation and calculations of eigenvalues and eigenvectors
RICCATI	Solves Riccati equations
RLOCUS	Plots the root locus for a continuous single-input/single-output system
RMS	Plots the root-mean square response of a continuous system
SPLIT	Splits a system matrix into its four individual matrices
ZEROS	Computes transmission zeros of a continuous or discrete state-space system
LYAPUNOV	Solves a continuous Liapunov equation
MARGIN	Computes gain and phase margins of a single-input/single-output continuous or discrete system
POLEPLACE	Calculates state feedback gains via pole placement for single-input continuous or discrete systems
REGULATOR	Computes the optimal state feedback gain for a continuous system
RESIDUES	Models residues of a continuous or discrete state-space system

Digital Signal Processing

The DSP module adds frequency analysis and filter design tools to $MATRIX_x$. Among the more important DSP functions are:

CORRELATE	Calculates the auto- and cross-correlation of data
DETREND	Removes the biases from the columns of a matrix
FFIR	Designs FIR filters, Hilbert transformers, and differentiators

FFT	Calculates the fast Fourier transform of input
FIIR	Designs low-pass, high-pass, and symmetric band reject IIR filters
FILP	Does filter propagation and simulation of discrete dynamic systems
FWIN	Designs FIR filters using the windowing method
HIST	Finds the histogram of a matrix
IFFT	Computes the inverse fast Fourier transform

Servo System Example

Fig. A.1 illustrates the interactions between $MATRIX_x$ and SYSTEMBUILD (the block diagrammatic $MATRIX_x$ simulation package). The problem posed is to meet a set of design specifications for the servo system shown in the figure by determining the value of the control gain, k, in the diamond-shaped block. The control gain must be specified so that the plant yields an output u with a fast rise time and minimal overshoot. The gain and compensator blocks must be implemented with a digital computer (sample/ hold running at 100 Hz).

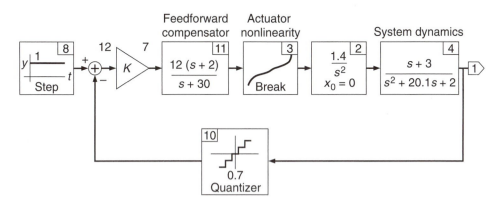

Figure A.1   Servo system using $MATRIX_x$ and SystemBuild.

The Approach

The design/analysis/simulation strategy is straightforward.

- Construct a model of the complete control system including nonlinearities with the graphical language of SystemBuild, the MATRIX$_x$ block diagrammatic simulation package. This is shown in Figure A.1.

- Use the forward loop model to study the frequency response (via Bode diagrams) of the open-loop system and the root locus of the closed-loop system as a function of control gain assuming, since these are linear tools, that the actuator and quantizer are represented by linear approximations.

- Based on these analyses, we choose a value for control gain of approximately 100. K=100 meets the specs of fast rise time and small overshoot.

- We then build the closed loop system of Figure A.1 with k=100. We include the actuator and quantizer nonlinearities and run simulations of the system using a step input as the reference signal. Outputs of the form shown in Fig. A.2 result. Note that the output indicates a limit cycle due to the nonlinearities in the loop. This may be acceptable, but if not, additional filtering may be required to further smooth the oscillations.

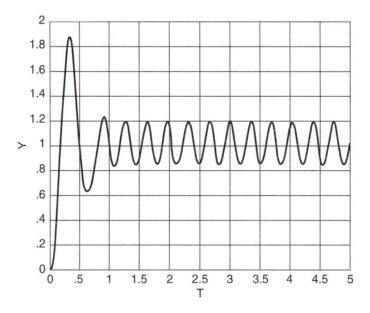

Figure A.2   Output for the servo system depicted in Figure A.1.

For more information on the MATRIX$_x$ software package and other products from Integrated Systems, Inc., contact Integrated Systems, Inc., 201 Moffet Park Drive, Sunnyvale, CA 94089; Phone: (408) 542-1500.

A.2 MATLAB[2]

MATLAB is an environment for performing technical computations. It provides engineers, scientists, and other technical professionals with a single interactive system that integrates numeric computation and scientific visualization. Its features include:

- Over 500 math and matrix functions
- Data analysis
- 2-D and 3-D visualization
- Linear algebra
- High-speed computational kernel
- High-level language for algorithm development
- Extensible via C and Fortran
- SIMULINK environment

SIMULINK

SIMULINK is an interactive software environment for modeling and simulating dynamic systems, including linear, nonlinear, discrete, continuous, and hybrid systems. Within this environment, block diagram models can be constructed with click-and-drag operations, changes to model parameters can be made on the fly, and simulation results can be displayed in real time fashion.

SIMULINK is built on top of the standard MATLAB computing environment, which includes a variety of toolboxes, like the Control System Toolbox and the Fuzzy Logic Toolbox. Included in the SIMULINK package are several libraries of components for modeling real world systems, e.g., in the nonlinear library are relays, backlash elements, dead zones, quantizers, and saturation elements. Combining these with linear systems blocks, we can construct block diagrams of highly nonlinear systems. Then it is possible to select appropriate inputs, run simulations, and observe responses. Control systems that are too complex to describe analytically can be designed in an empirical fashion using SIMULINK.

---

2   Section A.2 courtesy of MathWorks, Inc. Portions reprinted with permission.

Control System Toolbox

The Control System Toolbox is a collection of MATLAB functions for the analysis and design of automatic control systems. The central feature of the package are functions that implement the most useful "classical" transfer function and "modern" state-space control techniques.

The Control System Toolbox builds on MATLAB's extensive set of matrix and math functions to provide tools specialized to control engineering. The Control System Toolbox is a collection of algorithms that implement common control system design, analysis, and modeling techniques.

System models. The Control System Toolbox works with four different types of system models:

- State space
- Transfer function
- Pole zero gain
- Partial fraction (residue)

These models can be in continuous time or they can be in discrete time for digital control applications.

Model conversions. A set of conversion functions allows models to be converted between the various representations:

ss2tf	State space to transfer function
ss2zp	State space to zero pole
tf2ss	Transfer function to state space
zp2ss	Zero pole to state space
zp2tf	Zero pole to transfer function
residue	Transfer function to residue

Models can also be converted between continuous and discrete-time representations by a choice of methods:

c2dm	Continuous-to-discrete conversion
d2cm	Discrete-to-continuous conversion

Other functions provide tools for constructing larger and more complicated models out of basic system units:

append	Appends system dynamics
connect	Block-diagram modeling
parallel	Parallel system connection
series	Series system connection
feedback	Feedback system connection
reg	Forms optimal controller
estim	Forms Kalman estimator

Time response. Models may be simulated in the time domain. Functions for calculating time response include:

Impulse	Impulse response
Step	Step response
lsim	Simulation with arbitrary inputs

The discrete-time versions of these functions are called *dimpulse, dstep,* and *dlsim.* Options are provided for automatic plotting and time vector selection.

Frequency response. A group of functions calculates frequency-domain characterizations of models. Frequency response functions include:

bode	Bode plots
nyquist	Nyquist plots
nichols	Nichols plots
sigma	Singular-value plots

The discrete-time equivalents of these functions are *dbode, dnyquist, dnichols,* and *dsigma.* Options are provided for automatic plotting and frequency vector selection. In Fig. A.3 we present a typical Bode plot generated by the BODE statement.

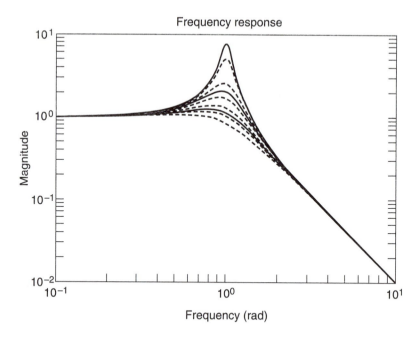

Figure A.3   Frequency Response.

Feedback Gain Selection. The Control System Toolbox contains tools for the selection of closed-loop feedback gains using both "classical" methods and "modern" optimal control methods. The classical methods include:

damp	Damping factors
margin	Gain and phase margins
place	Pole placement
rlocus	Auto-plotting root locus
rlocfind	Interactive root-locus gain determination

Functions for optimal control include:

lqe, dlqe	Linear-quadratic estimator design
lqr, dlqr	Linear-quadratic regulator design

These tools are used with the functions *reg* and *estim* for the design of regulators and Kalman estimators. In Fig. A.4 we present a typical root-locus plot in the z-plane.

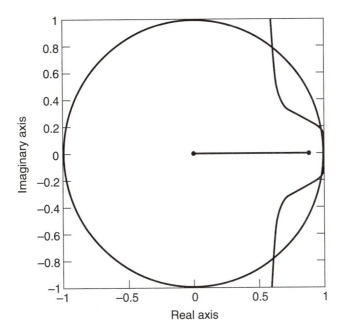

Figure A.4   A z-plane root-locus plot.

Model realizations.   Practical control systems are often designed around reduced-order models, after unimportant dynamics have been eliminated.   Functions for model order reduction and for transformation to related alternative realizations include:

ctrbf	Controllability staircase form
obsvf	Observability staircase form
minreal	Minimal realization and pole–zero cancellation
balreal	Balanced realization
modred	Model order reduction

Model Properties.   A collection of functions that calculate some model properties useful for certain advanced analyses:

gram	Controllability/observability gramian
ctrb	Controllability matrix
obsv	Observability matrix
tzero	Transmission zeros
lyap, dlyap	Liapunov equation
covar, dcovar	Covariance response

Signal Processing Toolbox

The Signal Processing Toolbox is a collection of MATLAB functions that provides a customizable framework for digital and analog signal processing. The Signal Processing Toolbox is not as important as the Control System Toolbox for the design of control systems, but since there is an occasional need for it, we include some of the features of the Signal Processing Toolbox. This toolbox includes functions for:

- Signals and linear system models
- Digital and analog filter design
- Filter analysis and implementation
- Transforms
- Statistical signal processing
- Signal, spectrum, and model visualization
- Parametric time-series modeling
- Waveform generation
- Windowing
- Specialized operations

The Signal Processing Toolbox provides an ideal environment for digital signal processing, system design, modeling, and algorithm development. Using filter design functions, for instance, we can design a 10th-order bandpass elliptic filter whose frequency response is indicated in Fig. A.5.

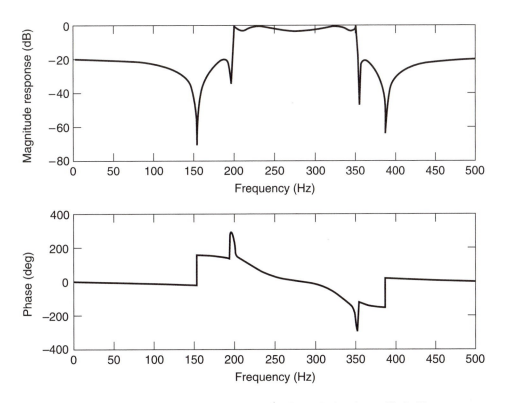

Figure A.5 Magnitude and phase response of a 10th-order bandpass elliptic filter.

Partial List of Functions

Signal and Linear System Models. The Signal Processing Toolbox provides numerous models for representing signals and linear time-invariant systems. The toolbox also provides functions for transforming models freely from one representation to another. Included are functions for:

- Time-domain (single channel and multichannel)
- Magnitude phase
- Laplace transform
- z-transform
- State-space form
- Lattice structure

- Zero–pole-gain form
- Second-order sections
- Partial-fraction expansions

The Signal Processing Toolbox also supports a full suite of design methods for finite impulse response (FIR) and infinite impulse response (IIR) digital filters.

Transforms.   Functions for computing the discrete Fourier transform and other transformations useful in analysis, coding, and filtering include:

fft	Fast Fourier transform
ifft	Inverse fast Fourier transform
fft2	Two-dimensional fast Fourier transform
ifft2	Two-dimensional inverse fast Fourier transform
fftshift	Swap halves of transformed vectors
hilbert	Hilbert transform
rceps	Real cepstrum and minimum phase reconstruction
cceps	Complex cepstrum
czt	Chirp z-transform

Statistical Signal Processing.   The following functions compute the power spectrum estimate and related functions, given one or two channels of data:

cohere	Frequency-domain coherence function
corrcoef	Correlation coefficients
csd	Cross spectrum estimation
cov	Covariance
psd	Power spectral density
tfe	Transfer function estimation (magnitude and phase)
xcorr, xcorr2	One- or two-dimensional cross-correlation function

Parametric Modeling.   Functions for computing parametric models of signals and linear systems from time-domain impulse response or frequency response data include:

invfreqs, invfreqz	Analog or digital filter fit to frequency response
levinson	Levinson's recursion time domain

lpc	Linear prediction (AR) coefficients
prony	Prony's discrete filter fit-to-time response
stmcb	Linear model using Steigliz–McBride method

Visualization. The toolbox provides a variety of graphical methods for visualizing time series, spectra, time-frequency information, and pole–zero locations, including:

stem	Discrete-time stem plots
strips	Strip plot of a signal on multiple lines
zplane	Pole–zero locations on the imaginary versus real plane

Signal Generation. The toolbox provides functions for generating synthetic waveforms beyond the standard trigonometric functions. User-defined functions can also be added. Signal generation functions include:

diric	Dirichlet function (periodic sinc)
sawtooth	Sawtooth and triangle-wave function
sinc	Sinc function $[\sin(\pi x / \pi x)]$
square	Square-wave function

In addition to the Control Toolbox and the Signal Processing Toolbox, MATLAB has recently made available a Fuzzy Logic Toolbox which has been useful in the design of Fuzzy Logic Controllers (FLCs).

Fuzzy Logic Toolbox

The Fuzzy Logic Toolbox for use within a MATLAB environment is a software tool for solving problems with fuzzy logic. This toolbox supports fuzzy clustering, Sugeno inferencing, and adaptive neuro-fuzzy learning. Although the toolbox has many interesting features, the most important for the needs of this book are the capabilities of creating and editing fuzzy inference systems (FIS's).

Three steps are involved in the FIS construction: fuzzification, inference, and defuzzification. Membership functions need to be created, fuzzy logic operators selected, and if-then rules applied. All rules are evaluated in parallel using fuzzy reasoning; then the results of the rules are combined and defuzzified to yield a crisp (nonfuzzy) output.

The easiest approach to FIS construction involves the use of the graphical user interface (GUI) tools that the toolbox provides. There are five basic GUI tools for FIS construction within the Fuzzy Logic Toolbox:

- The FIS Editor
- The Rule Editor
- The Membership Function Editor
- The Rule Viewer
- The Surface Viewer

For more information on the MATLAB and toolboxes from MathWorks, Inc., contact The MathWorks, Inc., 24 Prime Park Way, Natick, MA 01760-1500. Phone: (508) 653-1415, Fax: (508) 653-2997, E-mail: Info@mathworks.com.

dSPACE[1]

DSP-CITpro/eco Products from dSPACE

DSP-CITpro, developed by dSPACE, is an integrated software and hardware environment for rapid implementation of high-speed digital controllers, filters, hardware in the loop simulations and other high-end applications, using digital signal processors (DSPs) (Fig. B.1).

DSP-CITpro Software

IMPAC Package for implementation of linear and nonlinear systems consisting of the Implementation Expert IMPEX and at least one DSPL or C compiler.

IMPEX Program module for analysis and automatic implementation of linear time-invariant multivariable systems. DSPL source code generation for fixed-point processors and C source code generation for floating-point processors.

---

1 Appendix B is presented by courtesy of dSPACE GmbH. Portions reprinted with permission.

DSPL   The high level language DSPL provides special language Compiler
                    elements for the efficient programming of linear and nonlinear
                    multivariable systems on fixed-point processors. DSPL compilers are
                    available for Texas Instruments first and second Generation DSPs
                    (TMS 3201x and TMS 320C25).

NMAC   Nonlinear function table-lookup macro generators. Generated macros
                    can be included in a DSPL program by procedure calls. The NMAC
                    modules generate assembly source for Texas Instruments first and sec-
                    ond generation DSPs.

TRACE   Real-time trace modules for the individual DSP-CITpro/eco processor
                    boards. TRACE gives signal-oriented graphical insight into the DSP
                    program under real operating conditions. Interfaces to the analysis and

                    design tools MATLAB and MATRIX$_x$.

DSP-CITpro Hardware

The DSP-CITpro hardware line is a set of PC-AT compatible boards using the Texas Instruments TMS320 DSP family and consists of processor and peripheral boards for various applications. A separate high-speed bus (PHS-bus) is used for the communications between
the DSP and peripheral boards.

DS1001   Processor board with TMS320C25 fixed-point DSP (100ns cycle time).

DS1002   Processor board with TMS320C25 floating-point DSP (60ns cycle
                    time).

DS801   Switch for two processor boards to share a single set of peripheral
                    boards.

DS2001   ADC board: 5 parallel converters, 16 bit, 5 μs.

DS2002   ADC board: 2 parallel converters, 16 bit, 5 μs, 32 multiplexed chan-
                    nels.

DS2101   DAC board: 5 parallel converters, 12 bit, 3 μs.

DS3001   Incremental Encoder Interface board: 5 separate channels for direct
                    connection of standard incremental position sensors.

DS4001   32 bit I/O and 5 separate 16-bit timers (AM9513A) for various appli-
                    cations.

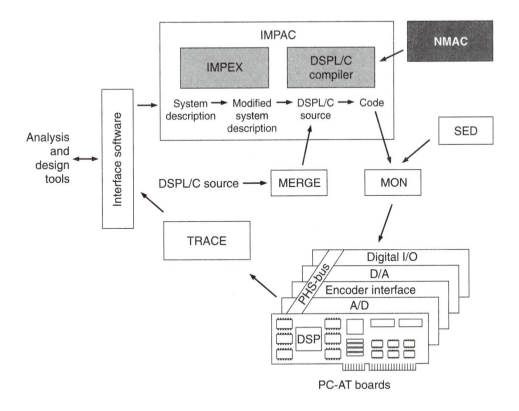

Figure B.1   DSP-CITpro Development System

DSP-CITeco Hardware

DS1101   Controller board with the TMS320E14 microcontroller DSP (160ns cy-
                cle time). Special on-board and on-chip peripherals are particularly
                suited to various applications, especially from the control field.

Expansion Boxes

PC-sized housings with enhanced power supply and cooling for large DSP-CITpro/eco
hardware systems. Interface boards and cabling to connect to the PC-AT host are included.

PX6, PX20   Expansion boxes with 6 and 20 slots (one slot used for box-side inter-
                 face board).

DS811   Separate host-side interface board for sharing an expansion box among
                 different PC-ATs.

Laboratory Equipment

EMPS300   High-precision, electromechanical positioning system for exper-
                 imentation with high-speed position control using DSP.

For more information on the products from dSPACE, Inc., contact dSPACE, Inc.,
22260 Haggerty Road – Suite 120, Northville, MI 48167, Phone: (248) 344-0096, Fax:
(248) 344-2060, or dSPACE GmbH, Technologiepark 25, D-33100 Paderborn, Germany
++49-5251-1638-0, Fax ++49-5251-66529.

Tables of Transforms

Table of z-Transforms

f(k)	F(z)
$\delta(k)$	1
$u(k)$	$\dfrac{z}{z-1}$
$ku(k)$	$\dfrac{z}{(z-1)^2}$
$k^2 u(k)$	$\dfrac{z(z+1)}{(z-1)^3}$
$k^3 u(k)$	$\dfrac{z(z^2+4z+1)}{(z-1)^4}$
$a^k u(k)$	$\dfrac{z}{z-a}$
$ka^k u(k)$	$\dfrac{az}{(z-a)^2}$
$ka^{k-1} u(k)$	$\dfrac{z}{(z-a)^2}$

$(k+1)a^k u(k)$	$\dfrac{z^2}{(z-a)^2}$
$k^2 a^k u(k)$	$\dfrac{az(z+a)}{(z-a)^3}$
$k^3 a^k u(k)$	$\dfrac{az(z^2+4az+a^2)}{(z-a)^4}$
$\dfrac{a^k}{k!} u(k)$	$e^{a/z}$
$\dfrac{1}{k} u(k)$	$\ln\left(\dfrac{z}{z-1}\right)$

These more general z-transform pairs are useful when the sampling T is not normalized to 1.0.

f(kT)	F(z)
$\sin k\omega T u(k)$	$\dfrac{z\sin\omega T}{z^2-2z\cos\omega T+1}$
$\cos k\omega T u(k)$	$\dfrac{z(z-\cos\omega T)}{z^2-2z\cos\omega T+1}$
$a^k \sin k\omega T u(k)$	$\dfrac{az\sin\omega T}{z^2-2az\cos\omega T+a^2}$
$a^k \cos k\omega T u(k)$	$\dfrac{z(z-a\cos\omega T)}{z^2-2az\cos\omega T+a^2}$
$e^{-akT} u(k)$	$\dfrac{z}{z-e^{-aT}}$
$(1-e^{-akT})u(k)$	$\dfrac{z(1-e^{-aT})}{(z-1)(z-e^{-aT})}$
$kTe^{-akT} u(k)$	$\dfrac{zTe^{-aT}}{(z-e^{-aT})^2}$

$(kT)^2 e^{-akT} u(k)$	$\dfrac{T^2 e^{-aT} z(z + e^{-aT})}{(z - e^{-aT})^3}$

Table of Laplace Transforms

f(t)	F(s)
$\delta(t)$	1
$u(t)$	$\dfrac{1}{s}$
$tu(t)$	$\dfrac{1}{s^2}$
$t^2 u(t)$	$\dfrac{2}{s^3}$
$t^n u(t)$	$\dfrac{n!}{s^{n+1}}$
$e^{-at} u(t)$	$\dfrac{1}{s + a}$
$te^{-at} u(t)$	$\dfrac{1}{(s + a)^2}$
$t^2 e^{-at} u(t)$	$\dfrac{2}{(s + a)^3}$
$t^n e^{-at} u(t)$	$\dfrac{n!}{(s + a)^{n+1}}$
$\cos \beta t u(t)$	$\dfrac{s}{s^2 + \beta^2}$
$\sin \beta t u(t)$	$\dfrac{\beta}{s^2 + \beta^2}$
$e^{-\alpha t} \cos \beta t u(t)$	$\dfrac{s + \alpha}{(s + \alpha)^2 + \beta^2}$

$e^{-\alpha t}\sin\beta t u(t)$	$\dfrac{\beta}{(s+\alpha)^2+\beta^2}$
$te^{-\alpha t}\cos\beta t u(t)$	$\dfrac{(s+\alpha)^2-\beta^2}{[(s+\alpha)^2+\beta^2]^2}$
$te^{-\alpha t}\sin\beta t u(t)$	$\dfrac{2\beta(s+\alpha)}{[(s+\alpha)^2+\beta^2]^2}$

Partial-Fraction Expansion Method

The partial-fraction expansion method is the easiest way of obtaining the inverse z-transform of a sequence. Consider the general z-transform form of a sequence f(k):

$$F(z) = \frac{b_0 z^n + b_1 z^{n-1} + \cdots + b_n}{z^n + a_1 z^{n-1} + \cdots + a_n} \qquad \text{(D.1)}$$

We must first factor the denominator polynomial to find the roots of the characteristics equation [poles of F(z)]. Then the transfer function in (D.1) can be represented as

$$F(z) = \frac{b_0 z^n + b_1 z^{n-1} + \cdots + b_n}{(z - p_1)(z - p_2) \cdots (z - p_n)} \qquad \text{(D.2)}$$

At this point we should consider the distinct poles and multiple poles of F(z) separately.

Distinct Poles

If all poles of F(z) are distinct, we can represent F(z) as

$$F(z) = \alpha_0 + \alpha_1 \left(\frac{z}{z - p_1} \right) + \alpha_2 \left(\frac{z}{z - p_2} \right) + \cdots + \alpha_n \left(\frac{z}{z - p_n} \right) \qquad \text{(D.3)}$$

The general expression for α_i is

$$\alpha_i = (\frac{z - p_i}{z})F(z)\Big|_{z=p_i} \text{ for } i = 1, 2, ...,n \qquad (D.4)$$

$$\alpha_0 = F(z)\Big|_{z=0} \qquad (D.5)$$

which means that we evaluate F(z) of (D.3) at z = 0.

Another way to interpret (D.4) is that we expand $\frac{F(z)}{z}$ and determine α_i just like we do in the familiar Laplace transform expansions. The general expression for f(k) is

$$f(k) = \alpha_0 \delta(k) + \alpha_1 (p_1)^k + \cdots \alpha_n (p_n)^k \quad \text{for } k= 0, 1, 2, ... \qquad (D.6)$$

Therefore, by using Equation D.6 we can decompose F(z) into a sum of simple terms, from which the sequence f(k) is derived in a term-by-term fashion.

Multiple Poles

If F(z) has multiple poles, then after the denominator of F(z) is factored, it will have the following form:

$$F(z) = \frac{b_0 z^n + b_1 z^{n-1} + \cdots + b_n}{(z - p_1)^{n_1}(z - p_2)^{n_2} \cdots (z - p_n)^{n_q}} \qquad (D.7)$$

where $n_1 + n_2 + \cdots + n_q = n$. The partial-fraction expansion of F(z) has the following form:

$$F(z) = \alpha_0 + \alpha_1 (\frac{z}{z - p_1}) + \alpha_2 (\frac{z^2}{(z - p_2)^2}) + \cdots + \alpha_{n_1} (\frac{\alpha_{n_1} z^{n_1}}{(z - p_n)^{n_1}}) \qquad (D.8)$$

$$+\beta_1 (\frac{z}{z - p_2}) + \beta_2 (\frac{z^2}{(z - p_2)^2}) + \cdots + \beta_{n_2} (\frac{z^{n_2}}{(z - p_2)^{n_2}})$$

$$+\cdots+\frac{\gamma_1 z}{(z-p_q)}+\cdots+\frac{\gamma_{n_q} z^{n_q}}{(z-p_q)^{n_q}}$$

where

$$\alpha_0 = F(z)\Big|_{z=0} \tag{D.9}$$

$$\alpha_{n_1} = (\frac{z-p_1}{z})^{n_1} F(z)\Big|_{z=p_i} \tag{D.10}$$

$$\beta_{n_2} = (\frac{z-p_2}{z})^{n_2} F(z)\Big|_{z=p_2} \tag{D.11}$$

$$\gamma_{n_q} = (\frac{z-p_q}{z})^{n_q} F(z)\Big|_{z=p_q} \tag{D.12}$$

REFERENCES

1. C. R. Wylie, Jr., *Advanced Engineering Mathematics*, McGraw-Hill New York, 1979.
2. J. A. Cadzow, *Discrete-Time Systems,* Prentice Hall, Upper Saddle River, NJ, 1973.
3. M. O'Flynn and E. M. Moriarty, *Linear Systems*, Wiley, New York, 1987.

Matrix Analysis

$\boldsymbol{A}$ssuming that the reader is already familiar with basic matrix algebra, such as addition and multiplication of matrices and calculating the determinant of a matrix, we present in this section several more properties of matrices that are useful in the analysis and design of digital systems.

Transpose of a Matrix

Consider the general form of an m × n matrix A:

$$A = \begin{bmatrix} a_{11} & a_{12} & \cdots & a_{1n} \\ a_{21} & a_{22} & \cdots & a_{2n} \\ \vdots & \vdots & \vdots & \vdots \\ a_{m1} & a_{m2} & \cdots & a_{mn} \end{bmatrix} \tag{E.1}$$

The transpose of A is obtained by replacing rows with columns:

$$A^T = \begin{bmatrix} a_{11} & a_{21} & \cdots & a_{m1} \\ a_{12} & a_{22} & \cdots & a_{m2} \\ \vdots & \vdots & \vdots & \vdots \\ a_{1n} & a_{2n} & \cdots & a_{mn} \end{bmatrix} \tag{E.2}$$

Rank of a Matrix

The rank of a matrix A is the maximum number of linearly independent column (or row) vectors contained within it.

Matrix Differentiation

The matrix differentiation is found by differentiating each element of the matrix. If

$$A = \begin{bmatrix} a_{11} & a_{12} \\ a_{21} & a_{22} \end{bmatrix} \tag{E.3}$$

then

$$\frac{dA}{dt} = \begin{bmatrix} \dfrac{da_{11}}{dt} & \dfrac{da_{12}}{dt} \\ \dfrac{da_{21}}{dt} & \dfrac{da_{22}}{dt} \end{bmatrix} \tag{E.4}$$

Matrix Integration

Matrix integration is found by integrating each element of the matrix. If

$$A = \begin{bmatrix} a_{11} & a_{12} \\ a_{21} & a_{22} \end{bmatrix} \tag{E.5}$$

then

$$\int A dt = \begin{bmatrix} \int a_{11} dt & \int a_{12} dt \\ \int a_{21} dt & \int a_{22} dt \end{bmatrix} \tag{E.6}$$

Exponential Matrix

An exponential matrix e^{At} is defined using the following series representation:

$$e^{At} = I + At + A^2 \frac{t^2}{2} + \cdots + A^k \frac{t^k}{k!} + \cdots \tag{E.7}$$

Properties of an exponential matrix:

1. $e^{At} \cdot e^{Bt} = e^{(A+B)t}$ (E.8)
    if and only if AB = BA.

2. $\dfrac{d}{dt} e^{At} = Ae^{At} = e^{At} A$ (E.9)

3. $e^{At} \cdot e^{-At} = I$ (E.10)
    where I is the identity matrix.

4. $[e^{At}]^{-1} = e^{-At}$ (E.11)

State Transition Matrix

The matrix $e^{A(t-\tau)}$ is called the *state transition matrix* and is usually denoted by $\phi(t)$. The following are some of the properties associated with the state transition matrix:

$$\frac{d\phi(t)}{dt} = A\phi(t), \ \phi(0) = I, \text{ if } \tau = 0 \tag{E.12}$$

$$\phi(t_2 - t_0) = \phi(t_2 - t_1)\phi(t_1 - t_0) \tag{E.13}$$

$$\phi(t_1 - t_2) = \phi^{-1}(t_2 - t_1) \tag{E.14}$$

Eigenvalue and Eigenvector

The eigenvalue of a square matrix A is a complex number λ for which the equation

$$Ax - \lambda x = 0 \tag{E.15}$$

admits a nontrivial solution. An eigenvector of a square matrix A is the vector x in the equation

$$(A-\lambda I)x = 0 \qquad\qquad (E.16)$$

with λ as the eigenvalue.

Hermitian Matrix

The Matrix $A_{n\times n}$ is Hermitian if $A^*=A$, where A^* is the complex conjugate transpose of A.

Symmetric Matrix

A Hermitian matrix A is symmetric if all elements of A are real.

Positive Definite

A symmetric matrix is positive definite if and only if its eigenvalues are all positive.

Motion Controller Boards[1]

F.1 PMAC-STD 32

STD 32 Intelligent Multiaxis Motion Controller

The PMAC-STD 32 (Fig. F.1 to F.3) is an intelligent programmable multiaxis controller (PMAC) for the STD 32 and STD bus. Based on a high-performance, digital signal processor (DSP) and gate array technology, the PMAC-STD 32 offers superior performance and programmability. It controls DC brush, DC brushless, AC induction, stepper and variable-reluctance motors.

The PMAC-STD 32 is a two-board set that contains a digital signal processor, four axes of motion control, as well as encoder feedback and ancillary control functions. An optional third board may be added to support a total of eight axes. The PMAC-STD 32 is a joint development of Delta Tau Data Systems and Ziatech. The board is based on the Motorola DSP56001 operating at 10 MIPS (millions of instructions per second). Its sophisticated motion control firmware uses a comprehensive series of functions that simplify many of the more complex motion control tasks:

- Motorola DSP 56001 digital signal processor
- Four or eight axes of motion control
- 20-MHz DSP clock (10 MIPS)

---

[1] Appendix F courtesy of Ziatech Corporation. Portions reprinted by permission.

- Simple commands for complex motion control
- STD 32 bus and/or RS-232/422 control
- Controls DC brush, DC brushless, AC induction, stepper, and variable-reluctance motors
- PID control loop with velocity and acceleration feedforward
- 10-MHz encoder rate
- 48-bit position range (approximately 64 million counts)
- 16-bit DAC output resolution
- 500-block-per-second execution rate
- S-curve acceleration and deceleration
- Linear and circular interpolation
- Cubic trajectory calculations, splines
- 60 microseconds per axis servo cycle time
- Watchdog timer
- 24 points of Opto 22 compatible digital 1/O
- G-code command processing for CNC
- Optional DOS-based executive for development

Functional Considerations

The PMAC-STD 32 combines powerful DSP technology, fourth-generation motion control firmware, and Delta Tau Data's custom gate array IC. PMAC-STD 32 supports four output channels and four feedback channels. This capability can be expanded to eight outputs and feedback channels with the addition of another stackable motion control card. This configuration allows the PMAC to command from one to eight axes simultaneously. Since each axis is completely independent, a single PMAC-STD 32 can command a single axis on each of eight different machines, eight axes of motion on one machine, or any combination in between.

DSP56001 Processor

The PMAC-STD 32 uses the Motorola DSP56001 running at 19.66 MHz. This digital signal processor (DSP) chip features 512 words of full-speed on-chip RAM memory, preprogrammed data ROMS, and special on-chip bootstrap hardware to permit convenient loading of user programs into the program RAM. An important feature for motion control applications is the chip's preprogrammed ROM that has a four-quadrant sine-wave lookup table that is used for phase commutation. The data arithmetic-logic unit (ALU), address logic units, and program controller operate in parallel. Therefore, an instruction prefetch, 24 x 24-bit multiplication, 56-bit addition, two data moves, and two address pointer updates can execute in a single instruction cycle.

Figure F.1   PMAC-STD 32.

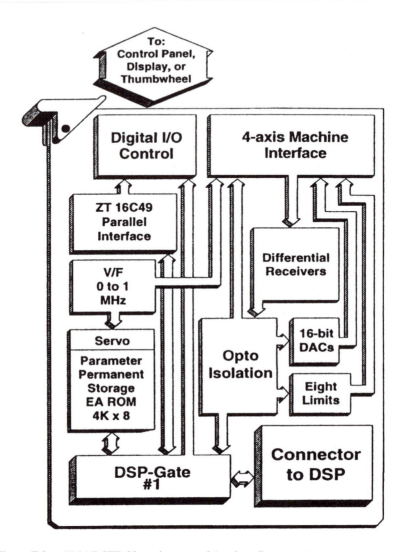

Figure F.2   PMAC-STD 32 motion control (top board).

PMAC Custom Gate Array IC

A key factor in the PMAC's performance is Delta Tau Data's 10,000-gate IC, which interfaces the DSP processor to the machine under control. The custom gate array includes:

- Four complete and identical encoder channels

- Four 16-bit, parallel/serial shift registers for receiving serial data from external A/D converters
- Four 16-bit, parallel/serial shift registers for sending data to the 16-bit D/A converters used to generate the output sequence
- 20 digital inputs used as encoder "C" channel, "home" flags, and "travel limits" and so on
- Eight general-purpose digital outputs

The encoder channel handles the highest possible speeds needed for machine control. Besides speed, the encoder channels provide the following special features:

- 10-MHz A/B quadrature encoder rate with 24-bit counters and registers
- Detection and reporting of illegal encoder transitions
- Digital noise filtering (necessary at high speeds)
- 1/T measurement, where time between encoder pulses is measured and used to calculate velocity (provides much smoother operation at low motor speeds)
- An externally triggered 24-bit position capture register that can be used to determine precise machine position upon occurrence of an external event
- A register that produces a pulse when the actual position equals the register's contents (eliminates the software delays usually associated with this type of output)
- A phase capture register for exact determination of the motor's phase commutation as a function of rotational angle
- A control register that allows the CPU to control all modes of operation directly, including encoder multiplication factor, direction control, YO polarity, and so on

Input Bandwidth

The PMAC-STD 32 is able to accept a 10-MHz maximum edge count rate with a digital filter. With the unique ability to use up to 5 bits of parallel binary data from the low-order, interpolated bits of high-resolution encoding devices, the PMAC can achieve effective input bandwidths of 320 MHz (10 MHz x 25). In a typical laser application with 0.1-microinch resolution, this translates to a velocity of 32 inches per second.

Feedback

The PMAC is designed to take incremental A/B quadrature encoder feedback without additional accessories. However, the appropriate accessories enable resolvers, potentiometers, absolute encoders, magnetostrictive linear displacement transducers, and laser interferometers to be interfaced to the PMAC. For calculation accuracy, and to avoid accumulated round off errors, the system uses a 24-bit up/down-position register. This hardware counter is automatically extended in software to 48 bits. A software parameter allows position rollover at a user-specified value: This is useful for controlling rotary axes.

Two methods are used by the PMAC to achieve subcount resolution with incremental encoding devices. The first is linear frequency response *decoding*. Each encoder channel has two timing registers. Register 1 holds T1, the time between the last two encoder transitions. Velocity is proportional to this term. T2, timer Register 2 records the interval since the last transition. Fractional distance traveled since the last transition is estimated by dividing T2 by Tl. This interpolation provides added smoothness to low speed moves. The second method of subcount resolution used by the PMAC is *parallel fractional feedback,* which accepts up to 5 bits of data in addition to the A/B quadrate edges. Data bits come directly from laser interferometers or as interpolated data from A/D conversion of analog encoders. Parallel fractional data are accepted both at rest and during programmed moves. If the A/D converter is unable to track the data during move sequences, the PMAC can change on the fly to 1/T subcount resolution.

Servo Loops

The PMAC-STD 32 provides both a PID and a notch position servo-loop filter with the following characteristics:

- A proportional-integral-derivative (PID) control loop with individually adjustable gain terms
- A tunable notch filter from 1 to 500 Hz, generally used to damp a critical machine vibration frequency, if required
- Velocity and acceleration feedforward terms for reduction or elimination of following error
- Double-position feedback loops, if required (one on the motor and one on the load), to control position accurately for a loosely coupled load

Also included is software support for "autotuning." It excites the load with impulse motions, evaluates and identifies the load characteristics and selects the optimal gain settings for the application.

Commutation

On-board commutation features of the PMAC provide support for the following motors:

- DC brushless or AC synchronous
- Variable (switched) reluctance
- AC induction
- Stepper motors

Commutation algorithms drive motor phases directly with only current loop bridges (transductance amplification) required in the amplifier. Two analog outputs per motor are required when the PMAC performs commutation, so a PMAC-STD 32-A2 with eight outputs can only control four motors if being used to provide commutation for all four motors. Using the PMAC to control commutation instead of the amplifier can significantly reduce amplifier cost and overall system cost. The torque performance and velocity smoothness of motors being commutated by the PMAC are also improved because the PMAC generates nearly perfect sinusoids.

Programming

The programming language used by the PMAC includes commands for diagnostics and setup, personalization, and on-line system controls. Diagnostics and setup commands allow the user to perform comprehensive checks of the hardware. It also provides verification of machine, control panel, and I/O operations. Autotuning and servo setup are available. There is also a *data gathering and display* mode that can store real-time data of position, following error, velocity, A/D input, D/A output, and so on. The data can then be presented in tabular or graphic form for analysis.

Personalization commands set the servo parameters and axis operational characteristics, such as following error limits, software travel limits, engineering units, backlash and deadband control, feed rates, encoder modes, commutation controls and other parameters that personalize each of the eight axes to their respective amplifiers, motors, and intended task. The PMAC can use any of the A/D converters as position feedback or feed rate control. A special section allows the storing of tabular data for functions that include lead screw error, backlash, tool radius and tool offset compensation.

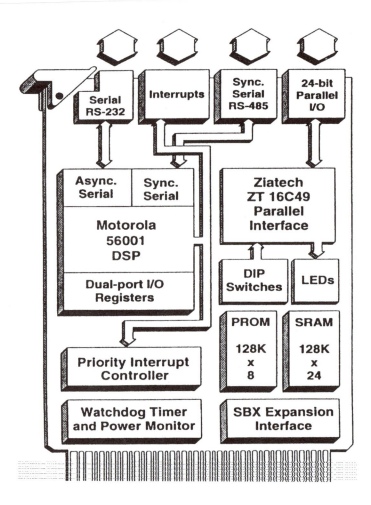

Figure F.3   PMAC-STD 32-digital signal processor (bottom board).

The on-line commands are a simple means to send instructions to the PMAC for immediate execution. Included are: axis select, jog ±, set zero position, start, stop, hold, quit, learn, and so on. The program buffer has special properties that enable the user to implement several working axis configurations. The eight axes can be independent of each other as if they are eight unrelated single-axis controllers, or conversely, configured as an eight-axis machine. This flexibility is accomplished by assigning a *coordinate system*. Within each coordinate system, one to three master axes can be assigned, with the rest assigned as slave axes. A coordinate system may contain one to eight axes, so one to eight coordinate systems may be assigned. Up to 256 programs may be stored in the PMAC, and any one may be executed by either of eight coordinate systems. This includes the case where one program is executed by eight coordinate systems synchronously or asynchronously.

A highly sophisticated DOS-based executive program is available and is strongly recommended for industrial systems. This program greatly facilitates the installation, startup, evaluation, servo-loop tuning, and programming of the PMAC-STD 32. A complete set of communications subroutines with source code is available from Delta Tau Data Systems for incorporation into host control programs as needed. Other programs are available which convert CAD drawings into PMAC programs. Training sessions are regularly conducted by Delta Tau Data Systems at its southern California headquarters or at customer sites. These classes provide valuable information and hands-on experience in a number of programming and hardware topics.

Background PLC Programs

While motion programs are running sequentially and synchronously in the foreground, the PMAC can run as many as 32 asynchronous PLC programs in the background. These programs perform many of the functions of a programmable logic controller while sharing the same logical contructs of motion programs, yet are unable to command the motion of axes. The PLC programs can monitor analog or digital inputs, set output values, send messages, monitor motion parameters, change gain values, and command motion Stop/Start sequences. PLC programs can even issue commands to the PMAC motion controller as if those commands came from the host processor itself. Typical PLC cycle times are 5 to 10 ms.

Daisy-Chaining Capability

The PMAC"'s software is limited to eight axes of control. Up to 16 PMACs may be daisy-chained, for a total of 128 axes of fully synchronized and interpolated controls with any type of amplifier and motor installed on any axis.

Interrupt Controller

An on-board interrupt controller allows the host computer to select, prioritize, and mask the PMAC-STD 32's eight interrupt sources. These interrupt sources include an external source and several internal sources such as PLC program-generated interrupts on velocity, acceleration rates, and so on.

F.2 STD/DSP SERIES

Motion Controller for Servo and Stepper Motors

The STD/DSP (Fig. F.4) controls both servos and steppers: A choice of PID, optional IIR, or fuzzy logic control with one board and one programming interface. Motor algorithms with velocity and acceleration feedforward types can even be mixed on a single board with precise control up to eight motors with update rates to STD/DSP's full range of control and I/O capability. 16-kHz outputs include 16-bit resolution analog signals for standard C function calls providing fast, familiar brush and brushless servomotors and high-resolution programming interface. Operating systems support instep and direction signals for open- and closed-loop: DOS, OS/20, ONX II 2.x and 4, Lynx/OS, and Miloop steppers. Inputs include incremental encoders, Microsoft Windows 3.1 and NT absolute encoders, analog sensors, laser interferometers, and Temposonics transducers (optional).

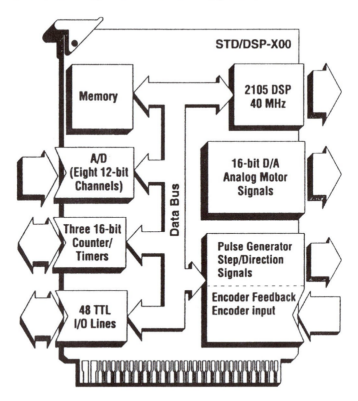

Figure F.4   STD/DSP series.

On-board controls:

- Brushless servos
- Brush servos
- Open-loop steppers
- Closed-loop steppers
- 2-, 4-, 6-, and 8-axis models

Standard C programming:

- Function libraries for DOS, Windows 3.1 and NT, ONX, Lynx/OS, and OS/2 (sources included)
- Interpolation: Two-axis circular, Eight-axis linear on independent coordinate systems
- Continuous contouring
- Up to 48 user I/O lines
- Opto 22 header trapezoidal, parabolic, and S-curve motion profiles
- Temposonics interface (optional) 1-MHz communication from host to on-board shared memory

Functional Considerations

The STD/DSP features advanced hardware capable of accomplishing the most complex and precise motion control applications.

High-Precision Motor Commands

The STD/DSP uses a true 16-bit D/A converter and command update times from 60 to 330 microseconds (.3 to 16 kHz) to provide extremely smooth, accurate motion. The accuracy of 0.08% at 1/2 RPS (equivalent to ±1 encoder count per DSP sample interval) was easily obtained with a standard Electrocraft BRU-200 servomotor with a 2000-line encoder. A 750-kHz voltage-to-frequency converter provides step and direction outputs in step rate increments of only 2 PPS for the fastest stepper accelerations possible.

Choice of Feedback Devices

The STD/DSP supports several position feedback devices with the standard firmware:

- Incremental encoder (differential or single-ended)
- Absolute encoder
- Laser interferometer (32-bit parallel)
- Analog sensor (LVDT or equivalent)

Device type can be software-selected and the selection stored on-board or set with C function calls

Analog Inputs

Eight channels of single-ended (four channels differential) analog inputs are provided on each board for analog position feedback, joysticks, tensioners, and so on. Either the DSP or the host processor can directly read the analog input values. Input voltage is 0 to 5 volts unipolar or ± 2.5 volts bipolar. Sample conversion rate is 75 kHz.

Servos and Steppers with One Board

Supporting servos and steppers with one board (in groups of two) minimizes programming time when a machine uses both servos and steppers when motor types are changed for new designs.

Optional Temposonics Support

A direct Temposonics interface can replace the step and direction outputs. This option sends a pulse to the transducer and gates the return pulse against a 40-MHz clock. The next pulse is sent automatically on return of the preceding pulse.

Dedicated and User I/O

The STD/DSP offers 72 bits of digital I/O to meet the requirements of most applications. Headers conform to Opto 22, Grayhill, and Gordos standards. Dedicated I/O for each axis includes positive/negative limits, home and amp fault inputs, and amp enable and in-position outputs. Additional dedicated inputs can be used to interrupt the DSP or the host CPU. Each board also provides 48 (two/four-axis models) or 24 bits (six/eight-axis models) of uncommitted user I/O. Note that user I/O pins are also used for interferometer and absolute encoder inputs when selected.

Torque Limiting

The maximum voltage (or step rate) output of the STD/DSP can be limited by a software command to a resolution of 16 bits.

Position Latching

When an input is sensed, the positions of all axes can be recorded in DSP memory in less than 1 microsecond per axis. The host can then read the positions at any time and reset the DSP for the next input.

Counter/Timer

The STD/DSP features a user-programmable three-channel 16-bit 82C54 counter-timer. The chip can be programmed to interrupt the host at set position intervals or time values.

Host Interrupts

The DSP can be instructed to interrupt the host on or prior to completion of a move, or at specified positions, times, velocities, and so on. The STD/DPS's flexible design makes simple tasks easy to program and simplifies programming for the most complex motion and I/O tasks.

Standard Languages

The STD/DSP controller is programmed through standard C language functions. The C libraries include over 250 functions with self-documenting names not arcane ASCII mnemonics. The use of standard languages from Borland, Microsoft, IBM, or ONX provides the user with a variety of tools and support not found in proprietary language controllers. All source code is provided as well as make files for several popular compilers. Utility functions are supplied for configuring the board, saving parameters to boot memory, and performing diagnostics. In addition, low-level functions are included for manipulating the DSP memory directly to perform special applications.

REFERENCE

1.   STD 32 *Technical Data Book,* Board and System Level Computer Products, Ziatech
      Corp., 1050 Southwood Drive, San Luis Obispo, CA 93401, Tel (805) 541-5088, Fax
      (805) 541-0488.

Sample DSP Programs

Assembly Code for the Torque Loop of Example 5.1[1]

Read- currents:	ADC-int:	
	i3=^I_ph;	{pointers for I_ ph CIRC array}
Read ADC registers	ml=l;	
and write to PARK input	13=3;	
registers	ax0=dm(ADCV);	{Read Iph(2) from ADC_ADMC200}
	dm(i3,ml)=ax0;	
	dm(PHIP1_VD)=ax0;	{write Iph(2) to PHIP1_ADMC200}
	axl=dm(ADCW);	{Read Iph(3) from ADCW_ADMC200}
	dm(i3,ml)=axl;	
	dm(PHIP2-VQ)=axl;	
	ay0=dm(theta)	{write theta to RHO_ADMC200:}
	dm(RHO)=ay0	{start RPARK by writing to ADMC200}
	i7 = ^ error_int	{error if next call takes > 37 cycles }
RPARK starts:	call over_current	{check for over current during PARK)
Clarke_Park	i7 = ^RPARK_int;	{interrupt vector set to RPARK }
	rti:	
Read PARK output	RPARK_int:	
registers including	ay0=dm(IX_PHV3);	{Read Iph(l) from ADCW_ADMC200}
derived third phase	dm(i3,ml)=ay0;	
current value.	I0 = ^lds	{pointer for Id and Iq}
	I0=2;	
set up pointers for	ay0=drn(ID_PHV1);	{Read Ids from ID_ADMC200}
current loop call	dm(i0,ml)=ay0;	

[1] Courtesy of Analog Devices, Inc. See Reference [1].

417

```
                          ax0=dm(IQ_PHV2);            {Read Iqs from IQ_ADMC200}
                          dm(i0,ml)=ay0;
                          I4=^Id_ref;
                          m4=1;
                          I4=0;
                          il=^  Did_n
                          I1=2;
                          call I_control             {current loop for Id return Vd in mr1}
                          dm(PHIP1_VD)=mrl;          {write Vd to VD_ADMC200}
                          il= ^Dlq n;
                          i4= ^ Iq-ref;

                          modify(i0,ml);             {i0 Flow points to lq}
                          call I_control             {current loop for Iq return Vq in mrl}
FPARK starts:             dm(PHIP2_VQ)=mrl;          {write Vq to VQ_ADMC200 axo=dm(tlieta)}
Park_Clarke-'            dm(RHOP)=ax0;              {write theta to RHO ADMC200: start FPARK}
                          i7 =^ error_int;           {error interrupt vector for > 37 cycles}
                          call ADC_ filters;
                          i7 = ^FPARK-int;           {store vector for park interrupt}
                          rti;

                          FPARK-int:
set up pointers           i2 =^AV_ph                 {pointers for V_ph};
for Vph                  m2=1;
                          12=3;
Read PARK output         ax0=dm(PHV1-VD);           {Read Va from PHV1_ADMC200}
registers                dm(i3,m3)=ax0;
                          ax0=dm(PHV2_VQ);           {Read Vb from PHV2_ADMC200}


                          dm(i3,m3)=ax0;
                          ax0=dm(PHV3);              {Read Vc from PHV3_ADMC200}
                          dm(i3,m3)=ax0;
                          il=^PWMTCHA;               {pointers for pwmtcha_ADMC200}
                          ml=l;
PWM_out                  Il=l;
                          ayl=pwmtdt;                {dead time compensations}
including:                axl=pwmtm_0;               {pwmtm/2}
deadtime adjustment      my0=dm{pwm sc};            {pwm scale factor}

                          rnr0=0;
                          cntr=3;
                          do pwmout until ce;
                          ar=axl+ayl,ax0=dm(i3,rn3); {t0 incl + tdt adjust}
                          af=pass ax0;               {check current polarity}
                          if It ar = axl-ayl;        {if –i then –tdt adjust}
```

```
                    mrl=ar;                      {load adjusted t0}
                    mx0=dm(i2,m2);               {v_ ph}
                                                 mr=mr+mx0*my0(ss);
                                                 {ton=t0+(+/-tdt)+v_ph*(t0/vbus) }
                                                 pwmout:

                    dm(il,ml)=mrl;               {write to pwmtchx_ADMC200}
                    call motion_control          {run position and velocity in remaining
                                                 time}

                    rt i;
```

current-loop{13 ops}	{	
PI error driven loop.	af = Diq_n1 = Iq_ref - Iq	
	mr = int = DIq_n *KIq	
	ar = sum = Diq_n1 + int	
	mr = DVq_n1 + int	
	ar = Vq_sum = DVq_n + (sum)*KPq	PI part
	mf = PHI = KEq+ld*Ls	
	mr = Vq=Vq_sum + velocity*PHI	machine equations
	}	
	I_control:	
	ax0=dm(i0,ml), ay0=pm(i4,m4);	{axo=lq ayo=lq_ref }
lq error	af=ay0-axo, mx0=dm@il,ml), my0=pm(i4,m4);	{mx0-DIq_n, my0=KIq}
	mr=mx0*my0(ss), my0=pm(i4,m4);	{mx0= KPq}
integral part:	ar = mrl+af, mrl=dm(il,ml)	{mrl=DVq_n}
proportional part:	mr=mr+af*my0(ss), ay0= pm(i4,m4);	{ay0=lRq}
save old values:	dm(il,ml)=af;	{Diq_n =af}
+machine equations	dm(il,ml)=mrl	{Dvq_n=mrl}
add IR drop to Vq	at =mrl+ay0, mrl= pm(i4,m4);	{mrl-Keq}
	mx0=dm(i0,ml), my0= pm(i4,m4);	{mx0=ld,my0 =Lsq}
q axis flux	mf=mr+mx0*my0(ss), mrl=ar;	
q back emf	mx0=dm(velocity);	
total Vq	mr=mr+mx0*mf(ss);	
	rts;	

DSP16 Code for the Adaptive Servo Controller of Example 5.3[2]

```
/*          DSP16 code for adaptive servo controller                      */
/*          The "set" mnemonic is optional.  It forces a single-cycle,    */
/*          single-word immediate load of YAAU registers for the DSP16A   */
/*          ("set" is default for the DSPl-@).  If the RAM variables are   */
/*          at addresses higher than 511, the "set" mnemonic should not    */
```

```
/*                  be used since the address will be truncated to 9 bits,      */
/*                  i.e., 511 maximum.
      auc=0x02              /* p <-- (x*y)*4        */
set  rl=mem5
set  r2=mem3
     pt=weight
     Y=1.0                      /* Input is step          */


                    *r2++=y          /* xn=1.0                  */
     do 40  {
                    y=a0    x=*pt++   /* Move initial weight    */
                    *rl++=x    /* vector from ROM to RAM  */
     }
     Y=0X0
     r3=meml
     do 78  {
         *r3++=y              /* Initialize RAM to zero            */
     }
         *r2++=y              /* Initialize mem4 to zero           */
         *r2++=y
         *rl++=y              /* Initialize tmpl & u(n)            */
         *rl++=y

start:
set  r0=mem4               /* points to u(n-1)        */
set  rl=mem2               /* points to c(n-2)       */
set  r2=meml               /* points to c(n-39)      */
set  r3=meml               /* points to c(n-39)      */
     pt=coef               /* points to state coef.  */
     i=-1                  /* pointer increment      */
     j=39
     k=-38
/* This section of the code computes the plant output */

                    y=*r0++      x=*pt++           /* y=u(n-1); x=0.004837)  */


          p=x*y      y=*r0++   x=*pt++              /* y=u(n-1); x=0.004678    */
a0=p      p=x*y      y=*rl++   x=*pt++              /*.y=cn(n-2); x= -0.9094   */
a0=a0+p   p=x*y               y=*rl++      x=*pt++            /* y=cn(n-1); x=1..9  */
a0=a0+p   p=x*y x=*r0++                    /* x=w40; prefetch for adap. Model  */
a0=a0+p                       y=*r2++      /* y=c(n-39); prefetch for adap. Model */
sdx=a0                                     /* Plant output via serial port*/
```

/* This section of the code computes the adaptive reference inverse model */

```
   p=x*y        *rl++=a0                    /* store output at cn  */
al=p            y=*rl++                     /*  y=xn(n-39)        */
   p=x*y        x=*r0++                     /*  x=w(39)           */
a0=p            y=*rl++                     /*  y=xn(n-38)        */


do 39 {
  p=x*y  *r2jk:y                            /*  move data         */
a0=a0+p   p=x*y x=*r0++                      /*  x=state coef.     */
al=al+p  y=*rl++                            /*  y=state variables */
                *r0++=a0                    /*  store plant input */
```

/* Compute reference plant output */

```
a0=y                                        /* save y            */
              y=a0    x=*pt++               /*retrieve y; x=0.01524 */


p=x*y              *rlzp:y        x=*pt++        /* x=0.0147      */
a0=p       p=x*y        y=*r0++      x=*Pt++     /* x=1.68476     */
a0=a0+p    p=x*y        *rozp:y      x=*pt++     /* x= -0.7147    */
a0=a0+p    p=x*y y=al                x=*pt++i
a0=a0+p                *r2 --
al=a0-y y=*r0                        x=*Pt++     /*  compute error  */
              y=al                   x=*Pt++


              P=x*y   y=*r2
al=p
j=-5
              *ro=al
              x=*r0++j


j=0
k=-l


do 40 {
  P=x*y        *r2jk:y                       /* update weight vector  */
               al=*r0
al=al+p        *rl++
               *r0--=al
}
                         *r0 --
                         *rl++
```

```
                                    y=*rl++
                                    *r0=y
                                    *rl=a0
end: goto start

/* state variables   */

ram
meml:   37*int              /* c(n-39),cn(n-38), ...,cn(n-3)  */
mem2:   43*int              /* en(n-2),cn(n-1),cn,xn(n-39),xn(n-38),..,xn(n-1)  */
mem3:   int                 /*  xn                                           */
mem4:   2*int               /*  un(n-1), un(n-2)                             */
mem5:   44*int              /*  w4O, w39, ..., wl, tmpl, un, r(n-1), r(n-2), tmp  */
.endram

/* fixed coefficients */

coef:   int         0.004837, 0.004678, -0.9094, 1.9
        int         0.01524, 0.0147, 1.6847,6, -.7147
        int         0.05


weight: int         -.000823,.000508,.001698,.002718      /* Initial weight value */
        int         .003550,.004199,.004694,.005092
        int         .005479,.005973,.006722,.007896
        int         .009681,.012263,.015807,.020418
        int         .026095,.032651,.039643,.046294
        int         .051458,.053657,.051201,.042415
        int         .025951,  .001145,  -.031634, -.070701
        int         -.112902, -.153631, -.187024, -.206343
        int         -.204497, -.174659, -.110923, -.008907
        int         .133761,.317147,.538981,.794912
```

REFERENCES

1. *AC Motor Control Using the ADMC200 Coprocessor,* Application Note, Analog Devices, Inc., Norwood, MA.
2. *Disk-Drive Servo Control with WEDSP16/DSP16A Digital Signal Processor,* Application Note, AT&T Corporation, June 1989.

Computer Architecture

The purpose of this appendix is to summarize the basics of computer architecture. The reader needs to have some basic understanding of hardware components within DSPs and other ICs used in modern control system design.

MEMORY

A memory unit is comprised of several equal-length registers that share a common set of inputs and outputs. Each memory unit has a limited capacity to store information, which is the maximum number of words it can store. The information stored in a memory register is a set of zeros and ones that can represent binary-coded information, an instruction code, or a number. The number of words (m) that can be stored and the number of bits (n) in each word always specify memories. There are n inputs to the memory unit and k select lines to select the words. READ and WRITE signals are two other inputs that let the information be read from or written into the registers. The n data output lines provide the information coming out of the memory. Fig. H.1 shows a block diagram of a memory unit.

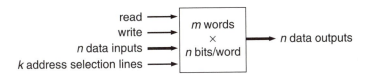

Figure H.1   A memory unit.

Based on their access time, memories can be classified to two major categories, random access memory (RAM) and sequential access memory (SAM):

1. *Sequential access memory (SAM).* A SAM is a memory device, and access to its contents is significantly dependent on the order in which the addresses are applied. Therefore, the information stored in the memory is accessible only at certain intervals of time, as in shift register memories.

2. *Random access memory (RAM).* A RAM is a general-purpose device that allows changes in its contents during the computational process. The access time in RAMs is always the same, so the memory contents can be accessed independent of the order in which addresses are applied. RAMs are used for applications where data are not constant and need to be updated.

Based on their write time, RAMs are classified as read-only memory (ROM), read-mostly memory (RMM), and read/write memory (RWM).

1. *Read-only memory (ROM).* ROMs are extensively used in the design of the combinational and sequential digital designs where the stored data do not need an update after production. The information stored in read-only memories is permanent; in other words, the contents of the memory are fixed during fabrication of the device and cannot be changed after manufacturing. ROMs are nonvolatile devices, therefore, their contents are not lost when the power supplied to the device is interrupted or removed. An m x n ROM is a memory unit, which contains m words that are n bits long each. The inputs to a ROM are k address lines that select one of the m words and output n data output lines, as shown in Fig. H.2.

Figure H.2   Block diagram of a ROM.

2. *Programmable Read-only memory (PROM).* PROMs are user-programmable ROMs. They originally contain fixed patterns, either all zeros or all ones. The desired program can be burned into the PROM by using PROM programmers. Once the PROM is programmed, the contents are fixed and cannot be changed.

3. *Erasable programmable Read-only memory (EPROM).* EPROMs are erasable programmable ROMs. As the name shows, the device is reprogrammable and reusable. EPROMs are either electronically or UV erasable using special tools. Although more expensive, EPROMs are used extensively in the early development of control programs because of their ability to be erased and accept the changes in the program.

Central Processing Unit

The central processing unit (CPU) is the basic building block of a computer that performs the instruction and data processing operations. The CPU is comprised of an arithmetic logic unit (ALU) for data processing and registers for data storage.

Arithmetic Logic Unit

The arithmetic logic unit (ALU) implements the basic data transformations in a microprocessor. The data transformations are the arithmetic and logic operations performed on one or two operands. The ALU contains an accumulator and a temporary register, which hold the operands. The result of the arithmetic or logic operations will be stored in an accumulator, replacing one of the operands. Some flip-flops are used as flags in the ALU to store the information that results from the arithmetic or logic operations. Some of the applications of flag registers are using a flag to show the divide-by-zero in a division operation, using a carry flag to show the carry-out in an addition operation and an overflow flag to show the overflow in arithmetic operations. The arithmetic operations that an ALU is capable of performing are addition and subtraction, and the logical operations are AND, OR, XOR, complement, rotate left or right. Fig. H.3 represents an ALU with its registers and flags.

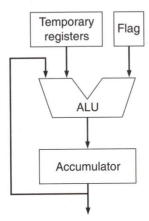

Figure H.3   Arithmetic logic unit.

Registers

Registers are groups of storage cells used in computers for binary information. Registers are made of flip-flops and some gates that control the transitions in flip-flops. The number of flip-flops equals the number of binary bits that can be stored in a register.

Shift Register

A shift register can shift the information stored in it either to left or right. An n-bit shift register consists of n flip-flops and some gates to control their transitions. Fig. H.4 shows the block diagram of a shift register.

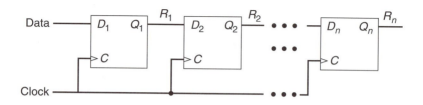

Figure H.4   Block diagram of a shift register.

Buses

System buses are parallel groups of wires that transfer the binary information among registers. A system bus is comprised of an address bus, a control bus, and a data bus.

1. *Address buses* are unidirectional and are groups of address signals that provide information to the device external to the microprocessor.

2. *Control buses* are unidirectional and are groups of control signals that are either input or output to the microprocessor.

3. *Data buses* are bidirectional and are groups of data signals that transmit information from microprocessor to external devices or from memory to microprocessor.

REFERENCES

1. M. Mano, *Computer System Architecture,* 2nd ed., Prentice Hall, Upper Saddle River, NJ, 1982.
2. K. L. Short, *Microprocessors and Programmed Logic,* Prentice Hall, Upper Saddle River, NJ, 1981.
3. C. H. Roth, *Fundamentals of Logic Design,* West Publishing, St. Paul, MN, 1979.

Index